Photosynthesis
Solar Energy for Life

Other Related Titles from World Scientific

Photosynthesis and Bioenergetics
edited by James Barber and Alexander V Ruban
ISBN: 978-981-3230-29-3

Molecular to Global Photosynthesis
edited by Mary D Archer and James Barber
ISBN: 978-1-86094-256-3

Solar Energy
edited by Gerard M Crawley
ISBN: 978-981-4689-49-6

Photosynthesis
Solar Energy for Life

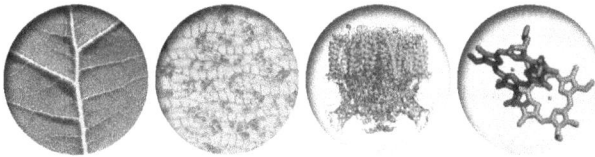

Dmitry Shevela
Umeå University, Sweden

Lars Olof Björn
Lund University, Sweden

Govindjee
University of Illinois at Urbana-Champaign, USA

World Scientific

NEW JERSEY · LONDON · SINGAPORE · BEIJING · SHANGHAI · HONG KONG · TAIPEI · CHENNAI · TOKYO

Published by

World Scientific Publishing Co. Pte. Ltd.

5 Toh Tuck Link, Singapore 596224

USA office: 27 Warren Street, Suite 401-402, Hackensack, NJ 07601

UK office: 57 Shelton Street, Covent Garden, London WC2H 9HE

Library of Congress Cataloging-in-Publication Data

Names: Shevela, Dmitry, 1979–   author. | Björn, Lars Olof, 1936–   author. |
    Govindjee, 1932–   author.
Title: Photosynthesis : solar energy for life / by Dmitry Shevela (Umeå University, Sweden),
    Lars Olof Björn (Lund University, Sweden),
    Govindjee (University of Illinois at Urbana-Champaign, USA).
Description: New Jersey : World Scientific, 2018. | Includes bibliographical references and index.
Identifiers: LCCN 2018044556 | ISBN 9789813223103 (hardcover : alk. paper)
Subjects: LCSH: Photosynthesis.
Classification: LCC QK882 .S49488 2018 | DDC 572/.46--dc23
LC record available at https://lccn.loc.gov/2018044556

British Library Cataloguing-in-Publication Data
A catalogue record for this book is available from the British Library.

Cover design: Dmitry Shevela (Umeå University).
Front cover background image: Sunset Over the Gulf of Mexico. Image credit: NASA/Terry Virts;
Editor: Sarah Loff. Image credits for front cover foreground photos: Dr Joanna Porankiewcz Asplund
(Agrisera, Sweden) for the picture of the leaf; and Dr Natalia Voronkina (Kaluga State University,
Russia) for the microscopy picture of cells in the Sphagnum leaf.

For any available supplementary material, please visit
https://www.worldscientific.com/worldscibooks/10.1142/10522#t=suppl

Typeset by Diacritech Technologies Pvt. Ltd.
Chennai - 600106, India

Contents

Preface

Jules Verne (1828–1905) wrote in 1874: *"Water will one day be employed as fuel—the hydrogen and oxygen that constitute it, used singly or together, will furnish an inexhaustible source of heat and light".*

Why should we focus on Photosynthesis, which, by using sunlight, becomes an *inexhaustible source* of energy for the planet? We hope that our book **"Photosynthesis: Solar energy for life"** will answer that for you. Photosynthesis is the most central and the most important process on the Earth, providing "order" from "disorder" using the all-abundant sunlight. Each year, photosynthesis fixes about 100 gigatons of carbon, supporting all life on our Planet, by providing us with oxygen to breathe and food to survive.

Our understanding of photosynthesis can now be said to have reached encyclopedic dimensions; there have been, in the past, many good books at various levels (Appendix A[1]). Our book is intended to fulfill the needs of undergraduate and beginning graduate students in all branches of biology as well as biochemistry, biophysics, and bioengineering. Photosynthesis will be central to future advances in producing more

---

[1] **Appendix A**

During 1945–1956, Eugene Rabinowitch (1898–1973), a post-doc student of the 1926 Nobel laureate, James Franck (1882–1964), wrote a 2088 page book in three physical volumes (Vol. I [1945], Vol. II [1951], Part 1, and Vol., Part 2 [1956]). It was indeed like a "Bible" of this field (available free at http://www.life.illinois.edu/govindjee/g/Books.html) at that time. If we were to write a similar book (almost impossible for most of us), it will take at least 30,000 pages! For college and university students, books by the following authors are worth mentioning: Robert Blankenship (*Molecular Mechanisms of Photosynthesis*, 2nd ed., 2014), Paul G. Falkowski & John A. Raven (*Aquatic Photosynthesis*, 2nd ed., 2007), Bacon Ke (*Photosynthesis*, 2003), David W. Lawlor (*Photosynthesis*, 2001), Eugene Rabinowitch & Govindjee (*Photosynthesis*, 1969; available free at http://www.life.illinois.edu/govindjee/g/Books.html.), and David Hall & Krishna K. Rao (*Photosynthesis*, 1967–1999). For advanced graduate and doctoral students, there are indeed many edited books by, e.g., Govindjee (*Bioenergetics*, 1975), Achim Trebst and Mordhay Avron (*Encyclopedia of Plant Physiology*, 1976), several volumes in "Topics in Photosynthesis", edited by James Barber (1976–1992), and more than 40 volumes in "Advances in Photosynthesis and Respiration" since 1994 (Govindjee and Thomas Sharkey, series editors; now Sharkey and Julian Eaton-Rye; see: http://www.springer.com/series/5599?detailsPage=titles).

food, more biomass, more fuel, and new chemicals for the expanding global human population. Further, the fundamentals of natural photosynthesis will provide the necessary information to engineer artificial photosynthesis, an important emerging field where chemists and engineers will play a major role.

Our book is presented in nine chapters arranged in a systematic pedagogical style with references so that the readers have an open-ended field before them. **Chapter 1** is an introduction explaining the "Why" of photosynthesis; in addition, it provides a historical background to the discoveries on which our current understanding of the process is based. The 'edifice of photosynthesis research' owes great debts to pioneers in the 17th, 18th and early 19th centuries, highlighted in this chapter. **Chapter 2** focuses on the nuts and bolts of the photosynthetic apparatus emphasizing the structures and the chemical components; thus, what makes photosynthesis *tick* is answered here. **Chapter 3** is devoted to the light-dependent reactions, the heart of the process of converting solar energy into biological energy. It is here that reader will find an in-depth discussion of the so-called Z-scheme of photosynthesis. **Chapter 4** completes the process of photosynthesis, i.e., the use of NADPH and ATP, the two major products of the light-dependent reactions, in producing carbohydrates by C3 (plants like rice) and C4 (plants like maize) carbon reduction pathways. The regulation needed for photosynthesis to perform optimally under different environments is addressed in **Chapter 5**, where we provide a concise picture of the regulation mechanisms that plants have developed and adapted to survive and excel under various conditions. These include the so-called "*state transitions*", where pigments move from one photosystem to another to balance light absorption for optimum overall photosynthesis, as well as "Non-photochemical Quenching (NPQ) of the excited state of chlorophyll *a*", one of the photoprotective mechanisms by which plants, algae, and cyanobacteria protect themselves in excess light. In **Chapter 6**, we discuss the impact of photosynthesis on the global environment and **Chapter 7** deals with anoxygenic photosynthesis by photosynthetic bacteria where water is not oxidized to oxygen, but instead other compounds such as H_2S are used as the primary electron source for these single photosystem organisms. **Chapter 8** deals with the fascinating story about the evolution of photosynthesis. **Chapter 9** ends the book featuring excitement around research in artificial photosynthesis. Finally, we remind the readers that we must move forward on all fronts to solve global challenges, using the knowledge of photosynthesis. We must not be afraid to make mistakes. As Jules Verne (in *A Journey to the Center of the Earth*) said, "*Science, my lad, is made up of mistakes, but they are mistakes which it is useful to make, because they lead little by little to the truth*". Let there be light and let there be hope. We hope that the readers will enjoy and learn from our book and work to improve our World.

Dmitry Shevela
(dmitry.shevela@umu.se)
Department of Chemistry, Chemical Biological Centre,
Umeå University, SE-90187 Umeå, Sweden

Lars Olof Björn
(Lars_Olof.Bjorn@biol.lu.se)
Department of Biology, Molecular Cell Biology, Lund
University, Sölvegatan 35, SE-22362 Lund, Sweden

Govindjee
(gov@illinois.edu)
Department of Plant Biology, Department of Biochemistry, and Center of Biophysics & Quantitative Biology, University of Illinois at Urbana-Champaign, Urbana, IL 61801, USA

Acknowledgements

We are grateful to Prof. Robert Blankenship (Washington University in Saint Louis, Missouri, USA) for reading the manuscript of our book, and for his valuable comments. We are indebted to Dr. Vandana Chakravartty for checking the entire book for language, and for her suggestions. We thank Prof. Donald Ort (University of Illinois at Urbana-Champaign, Illinois, USA) for reading and editing our Preface. We are obliged to our institutions for their support (Umeå University, Lund University, and University of Illinois at Urbana-Champaign). We are highly thankful to the following for providing us specific photographs, used and cited inside our book: Prof. Emeritus Andrew Staehelin (University of Colorado at Boulder, USA), Prof. J. William Schopf (University of California at Los Angeles, USA), Dr. Natalia Voronkina (Kaluga State University, Russia), and Dr. Joanna Porankiewicz-Asplund (Agrisera, Sweden). We are equally thankful to Joy Quek of World Scientific Publishing Company for her wonderful help in producing this book. Finally, and most importantly, we express our appreciation to our wives for their constant support during the last 2 years when we were working on this book: Lars Olof thanks Gunvor Björn, Dmitry thanks Izabella Surowiec, and Govindjee thanks Rajni Govindjee.

About the Authors

Dmitry Shevela obtained his BSc (Biology and Chemistry) in 2002 from the Kaluga State University, Russia. After exploring aspects of *"Bicarbonate effects on photosynthetic water oxidation"* at the Institute of Fundamental Biological Problems (Pushchino, Russia), he focused, at the Max Planck Institute for Chemical Energy Conversion (Mülheim an der Ruhr, Germany), on *"The role of inorganic cofactors in photosynthetic water oxidation."* For this work, done under the supervision of Vyacheslav Klimov and Johannes Messinger, he received, his PhD (in Biophysical Chemistry) in 2008, from the Technical University of Berlin (Germany). During his postdoctoral research at the Umeå University (Sweden), he studied photochemistry, leading to oxidation of water to oxygen, both in natural and artificial systems, and then at the University of Stavanger (Norway), he worked on the assembly of photosynthetic complexes. Currently, he is engaged in research, as a Senior Research Engineer, in the Department of Chemistry, at the Umeå University; he studies functional development and photoprotection mechanism of photosynthetic complexes during chloroplast biogenesis. In addition to his research, Dmitry is well known for being a successful scientific graphic designer. He is founder of graphic design company: ShevelaDesign AB (Sweden).

Lars Olof Björn did his undergraduate studies in Chemistry, Physics, Mathematics, and Botany, at the University of Lund (Sweden) and at the University of California Berkeley (USA). He obtained his PhD on *"The Effect of Light on Root Plastids"* under Hans Burström (founder of *Physiologia Plantarum*, which Lars Olof later helped to edit). Since then, he has been doing research on several aspects of photobiology. He is author of a highly popular book *"Light and Life,"* which was published in five languages: Swedish, English, German, Italian, and Japanese. He has also written popular science articles in Swedish (many published in *"Forskning och Framsteg"* and *"Svensk Botanisk Tidskrift"*). Lars Olof has served as a Professor of Botany at the Lund University, for almost 30 years (1972–2001), and he has been on the board of Association Internationale de Photobiologie (now International Union of Photobiology) for 12 years.

In addition, he has taught at the Agricultural University of Denmark, in Copenhagen, and at the South China Normal University, in Guangzhou, and has also been involved in research at the Carnegie Institution of Washington (Stanford, California), at RIKEN (Wako-Shi, Japan) and at the Abisko Scientific Research Station (Abisko, Sweden). He is now Professor Emeritus of Botany at Lund University. He is a member of the Royal Swedish Academy of Sciences.

Govindjee[2] (http://www.life.illinois.edu/govindjee/), known as Mr. Photosynthesis, studied Plant Physiology at Allahabad University, and Biophysics (with Chemistry and Physics) at the University of Illinois at Urbana-Champaign (UIUC). At UIUC, he obtained his PhD, in 1960, under Eugene Rabinowitch; he then served on the faculty at the UIUC for 40 years, retiring in 1999; since then, he has been Professor Emeritus of Plant Biology, Biochemistry, and Biophysics. Govindjee is the founding editor of the Series "*Advances in Photosynthesis and Respiration*," and of the Historical Corner of "*Photosynthesis Research*." He is an authority on the Primary Reactions of Photosynthesis, and is known for the discovery of the unique role of bicarbonate on the electron acceptor side of Photosystem II. His love of teaching is well known as he explains photosynthesis through a drama, where the students play the roles of different molecules. He is author of a world-renowned 1969 book "*Photosynthesis*," and co-editor of more than a dozen books. Govindjee has received many honors: 1981–1982 President of the American Society of Photobiology, recipient of the Lifetime Achievement Awards of the Rebeiz Foundation (2006), and of the UIUC (2008), and the 2007 Communications Award of the International Society of Photosynthesis Research. He is a fellow of the American Association of Advancement of Science, of the National Academy of Sciences of India, and of the National Academy of Agricultural Sciences of India.

---

[2] Govindjee has used one name only in all his publications and will continue to do so; although starting June 11, 2018, his legal name is: Govindjee Govindjee.

Chapter 1

Introduction

1.1 Why Study Photosynthesis?

At the dawn of photosynthesis research, the driving force was pure curiosity; it was truly a basic science. Today, we face several problems, which can only be solved by increased knowledge of climate change, energy demand, food supply, pollution, and much more!

Plants sequester the greenhouse gas carbon dioxide (CO_2) very efficiently from air. Can we learn from them and develop ways to clean the atmosphere more efficiently?

We have many questions before us…

How will plant behavior and plant photosynthesis change when the climate changes? How will these changes in photosynthesis provide feedback to the climate?

All food is produced directly or indirectly by photosynthesis. Can photosynthesis be increased to produce even more? Can it be modified to produce healthier food?

Mineral oil (from which gasoline/petroleum is produced) is a limited resource that was, ultimately, produced by photosynthesis long ago. In addition to serving as fuel, it is used to produce a number of chemicals, plastics, and other commodities. Can we modify photosynthetic organisms (plants, algae or bacteria) to produce the necessary raw materials for us?

Can we replace fossil fuel as an energy source with the products of *artificial photosynthesis*? Can electricity be produced directly by photosynthesis-inspired methods?

An increasing number of planets are being discovered outside of our own solar system. Do any of them harbor life? If there is photosynthetic

Photosynthesis: Solar Energy for Life by Dmitry Shevela, Lars Olof Björn and Govindjee
© 2018, published by World Scientific Publishing Co. Pte. Ltd. ISBN: 978-981-3223-10-3.

life out there, can we detect it from the spectra of their light-catching pigments, or from the gas that they produce? In addition, we ask if there are alternatives to the terrestrial compounds involved in photosynthesis?

1.2 History

1.2.1 *The discovery of photosynthesis*

Contrary to the opinion of the Greek philosopher Aristotle (384–322 B.C.), the idea that much of the substance of a plant was derived from water can be traced to a 2nd century A.D. unidentified author, referred to as Pseudo-Clement [Howe, 1965]. The original text of his "Recognition" is now lost, and survives only in a Latin translation made by Rufinus, Presbyter of Aquileia, in the 4th century A.D.

The notion of water being an element remained for a long time. Nicholas of Cusa (1401–1464), a German philosopher, scientist and a Cardinal, wrote, in clear contradiction to Aristotle, in the book *De Staticus Experimentis* [1450; cf. Krikorian and Steward, 1968] that if one puts earth (soil) and a plant (or seeds) in a pot, and allows the plant to grow hundred-fold in size, one would find that the soil would have lost very little weight, and the plant would have gained all its weight from the added water. This book might have inspired the physician Jan Baptista van Helmont (1580–1644) in Brussels, Belgium, to conduct his famous experiment [van Helmont, 1648]:

> I took an earthen pot and in it placed 200 pounds of earth which had been dried out in an oven. This I moistened with rainwater, and in it planted a shoot of willow [tree]which weighed five pounds. When five years had passed the tree, which grew from it, weighed 169 pounds and about three ounces. The earthen pot was wetted whenever it was necessary with rain or distilled water only. It was very large, and was sunk in the ground, and had a tin plated iron lid with many holes punched in it, which covered the edge of the pot to keep air-borne dust from mixing with the earth. I did not keep record of the weight of the leaves, which fell in each of the four autumns. Finally I dried out the earth in the pot once more, and found the same 200 pounds, less about 2 ounces. Thus, 164 pounds of wood, bark, and roots had arisen from water alone.

Thus, van Helmont did not understand that a large part of the plant substance is derived from the air, which was still considered to be an individual element, just as water. The notion that air contains a gas that is important for life can be traced back to Michael Servetus (~1510–1553), who published the book *"Christianismi Restitutio"* which describes how blood, when passing through the lungs, changes color. The book was disliked by the authorities in Geneva (Switzerland), so both Servetus and most copies of his book were burnt; only one copy seems to have survived. In 1668, John Mayow, a former assistant to Robert Boyle [Severinghaus, 2014], published a book in which he showed that air was a mixture of two components of which one-fifth was essential for life; it is consumed during breathing, and in fueling fire, and thus provides both body heat and energy.

In 1771, Carl Wilhelm Scheele (1742–1786) from the then Swedish province of Pomeronia in what is now northern Germany, while working at Uppsala University, produced a gas by heating mercury oxide. He called it *"fire air"* since it supports fire even better than air. Because of several unfortunate circumstances, his findings were not published until 1777. By then Joseph Priestley (1733–1804), of England, had already replicated his earlier findings, giving the produced gas the name *"oxigen"* ("oxi" meaning acid), and discovered several other gases. On August 17, 1771, he placed a sprig of mint into the "noxious air" produced by a lit candle, in an airtight container. Ten days later, a candle burned in the air perfectly well. He carried out the experiment 10 times during the summer of 1771 and repeated this experiment in the summer of 1772. *The best results were obtained when spinach was used, and this plant has remained a favorite material for photosynthesis researchers.* On March 8, 1774, Priestley decided to test the effects of oxygen on a mouse. He found that a mouse could survive in "fouled air" after it had been "dephlogistonated" [dephlogisticated] by a green plant. Although this is commonly regarded as the discovery of photosynthesis, Priestley did not notice the importance of light or CO_2 (the latter had been discovered by the Scottish chemist Joseph Black [1728–1799], already in 1755). Surprisingly for us in the 21st century, Priestley continued to believe in the *phlogiston* theory for the rest of his life. It was the French chemist Antoine Laurent de Lavoisier (1743–1794) who disposed of the old ideas of air and water being elements by showing that water is made up of hydrogen and oxygen. Methods improve, and the death of old ideas becomes less painful than that of Michael Servetus, since, instead of being burnt alive, Antoine Laurent de Lavoisier was executed by the guillotine! Lavoisier had angered Jean Paul Marat (1743–1793), another scientist whose writings he had criticized. Marat, in turn, was stabbed to death in his bathtub by Charlotte Corday. During this time, the situation for Priestley grew very turbulent and he was forced to flee to America (USA).

In 1778, the Dutch physician Jan Ingenhousz (1730–1799) repeated Priestley's experiments in a rented villa in England. He kept some jars and plants in darkness, and exposed others to sunlight. He found that a candle would burn longer, and a mouse would be revived only if the plant was exposed to sunlight. He clearly discovered the role of light in photosynthesis. In 1782, the Swiss botanist Jean Senebier (1742–1809), while modifying Ingenhousz's experiments, proved that plants absorb CO_2.

Nicolas Théodore de Saussure (1767–1845) extended, in an admirably quantitative way, Senebier's experiments. de Saussure described how plants quickly became yellow and wilted in light after he removed CO_2 from the air, which he did by using calcium hydroxide. Further, he noticed that plants evolved oxygen (O_2) only when CO_2 was present in the air surrounding them [de Saussure, 1804]. However, he drew the erroneous conclusion that CO_2 was decomposed so that free O_2 was released from it. In an experiment with periwinkle (*Vinca minor*), de Saussure found that a sample of air before the experiment contained 4199 cm^3 nitrogen (N_2), 1116 cm^3 O_2, and 431 cm^3 CO_2, i.e., a total of 5746 cm^3 gas. In another sample of 5746 cm^3 of air in which he had kept plants illuminated for some time, he found 4338 cm^3 of N_2, 1408 cm^3 of O_2, and no CO_2. The increase of O_2 was 292 cm^3, which was 139 cm^3 less than the amount of CO_2 that had disappeared. de Saussure, once more, drew another erroneous conclusion that some CO_2 had been transformed to N_2. This example shows that experimental science is not always as straightforward as one would like to think. We have described this example here, since it is an early example of the use of a unit still in use in science (de Saussure called it "*centim. cub.*"). In other experiments, de Saussure showed that plants increased their content of carbon, just as CO_2 disappeared. He carried out experiments with many different plants, always trying to work as quantitatively as possible, and concluded that the amount of carbon gained matched what was obtained from the air, and that, together with water, gave rise to the plant material.

Ingenhouz in "*An essay on the food of plants and the renovation of soils*" (1796) translated the whole process of photosynthesis from the old phlogiston concept to the new chemistry of Lavoisier, in a formal language that is still used today:

$$CO_2 + H_2O + light \rightarrow O_2 + plant\ matter \qquad (1.1)$$

For basics, see Rabinowitch and Govindjee [1969] and Blankenship [2014] and for a complete discussion of the evolution of the equations and the models of photosynthesis since 1840 until 1960, see Nickelsen [2015].

1.2.2 The concept of energy

The concept of "*energy*" had not been invented even at the beginning of the 19th century. Often Julius Robert von Mayer (1814–1878), from Germany, is given credit for it, but it is closer to the truth to say that several scientists contributed to its gradual emergence [Lopes Coelho, 2009; Cahan, 2012; Kipnis, 2014]. This remained a point of contention among scientists of the 19th century, and even today historians argue about this. Since we now understand photosynthesis to be a transformation of light into chemical energy (or life energy, bioenergy) we shall expand a little further on this.

This story begins with Nicolas Léonard Sadi Carnot (1796–1832), a French military engineer. The concept of "*work*" as a product of force and displacement was already there. Inspired by the development of steam engines, Carnot started to think about the relation between heat and work, and devised his famous "Carnot cycle" (in the "*Reflections on the Motive Power of Fire*" [1824]). The terms energy and entropy were not yet coined, and Carnot had only a fuzzy concept for them. If heat was a form of energy, why was the presence of something cold necessary to extract work from heat? As a military engineer, Carnot was not only familiar with how bullets and cannon balls were set in motion by gunpowder, but he also had ideas about chemical energy.

In 1841, von Mayer submitted his first scientific paper "*Über die quantitative und qualitative Bestimmung der Kräfte,*" to Annalen der Physik und Chemie. Being a medical doctor he did not use the terminology of physics of the day and expressed his ideas in a rather vague manner. His paper was rejected and not even returned to him. In 1842, he succeeded to have another paper published with the title *"Bemerkungen über die Kräfte der unbelebten Natur"* ("Remarks on the forces of the non-living nature") in Justus Liebigs Annalen der Chemie [Mayer, 1842]. In this paper, he gave an estimate of the amount of work corresponding to a certain amount of heat; this work was not well received at that time. However, in 1845, he gave an elaborate account in *"Die organische Bewegung in ihrem Zusammenhange mit dem Stoffwechsel"* ("Organic motion in its connection to metabolism") [Mayer, 1845]. In this work, he had formulated for the first time the *law of conservation of energy* (the first law of thermodynamics). Further, he extended this principle also to the living world and expressed it as:

> Die einzige Ursache der tierischen Wärme ein chemischer Prozess (The only reason for animal heat is a chemical process).

In 1867, he wrote in *"Die Mechanik der Wärme"* ("The mechanics of heat"):

> Nature has put itself the problem how to catch in flight light streaming to the earth and to store the most elusive of all forces in a rigid form. To achieve this aim, it has covered the crust of earth with organisms, which in their life processes absorb the light of the sun and use this power to produce a continuously accumulating chemical difference.

In German, the original text was:

> Die Natur hat sich die Aufgabe gestellt, das der Erde zuströmende Licht im Fluge zu erhaschen, und die beweglichste aller Kräfte, in starre Form umgewandelt, aufzuspeichern. Zur Erreichung dieses Zweckes hat sie die Erdkruste mit Organismen überzogen, welche lebend das Sonnenlicht in sich aufnehmen und unter Verwendung dieser Kraft eine fortlaufende Summe chemischer Differenzen erzeugen in die organische Bewegung in ihrem Zusammenhange mit dem Stoffwechsel.
> — Julius Robert von Mayer [1867]. *Die Mechanik der Wärme in gesammelten Schriften.* Cotta, pp. 53–54.

Where von Mayer uses the word *"forces"* we would use *"energy"*, and for *"chemical difference"* we would use *"chemical energy"*.

In 1850, Rudolph Clausius (1822–1888) also formulated the first law of thermodynamics:

> In all cases in which work is produced by the agency of heat, a quantity of heat is consumed which is proportional to the work done; and conversely, by the expenditure of an equal quantity of work an equal quantity of heat is produced.

Independently of von Mayer, the law of conservation of energy was also formulated in 1843 by James Prescott Joule (1818-1889) in England and by Ludvig August Colding (1815-1888) in Denmark. Joule was the first one to establish an almost correct equivalence between work and heat. In 1847, Hermann von Helmholtz (1821–1894) summed up the various versions of the first law of thermodynamics. Although the term "*energy*" had already been in use by the ancient Greeks, and it was just about to be established at that time, yet von Helmholtz still used "*Kraft*" instead, equivalent to what we call "*force*". Further, von Helmholz defined the concept of entropy and introduced the concept of free energy as opposed to total energy [Helmholtz, 1847].

At that time, it was not clearly understood that heat is a form of motion (of atoms and molecules), even though the idea had been vaguely around for a long time. It was left to Ludwig Eduard Boltzmann (1844-1906), from Austria, and James Clerk Maxwell (1831-1879), from Scotland, to formulate this precisely. Maxwell should also be added to the list of 19th century physicists, for having contributed to our understanding of photosynthesis. In 1865, Maxwell had published "*Dynamical Theory of the Electromagnetic Field*" where he already had his famous set of equations in this area [Maxwell, 1865]. There, he showed that light is an electromagnetic phenomenon. He carried out important research in various branches of science, especially physics. We agree with the view of Albert Einstein (1879–1955) that until the end of the 19th century, Maxwell and Newton (see below) were the most brilliant scientists that ever lived.

1.2.3 *Early research on photosynthesis*

1.2.3.1 *What absorbs the light for photosynthesis?*

The next important step forward was by Theodor Wilhelm Engelmann (1843–1909), who had invented an extremely sensitive method ("biosensor") for showing the presence of oxygen [Engelmann 1882]. He used bacteria that actively swim toward high O_2 concentration, and he watched them using a microscope. When he projected a spectrum of visible light on a filament of a green alga, he found that most bacteria were at places, which were illuminated by red or by blue light (Fig. 1.1), while fewer bacteria were under green light. He knew that chlorophyll (Chl) absorbs red and blue light more strongly than green light; he drew the important conclusion that it is Chl that absorbs the light resulting in oxygen evolution. By his experiment he also introduced *action spectroscopy* (measurement of action spectra), which has proven important for the understanding of many other processes in which light interacts with the living matter, and which has been important in the further elucidation of the photosynthesis process.

Richard Martin Willstätter (1872–1942) and his coworkers elucidated the chemical structure of Chl, and showed that various land plants contain a mixture of two types of Chl: Chl *a* and Chl *b*. Willstätter was awarded the 1915 Nobel Prize in chemistry in recognition of his work on chlorophylls and anthocyanins.

In the early 20th century Chl was seen as a molecule, which, upon absorption of light, could split CO_2 into free O_2 and "active carbon" that could then unite with water to form a carbohydrate (which is the reason we use the term "carbohydrate"). There was a belief that once the Chl, the substance "catching" the sunlight, has been understood, we would also understand the process of photosynthesis. As we shall see in the following chapters, there is a lot more than the involvement of Chl in this wonderful process.

1.2.3.2   What goes in, and what comes out?

As mentioned above the basic overall reaction of oxygenic photosynthesis is:

$$CO_2 + H_2O + \text{light} \rightarrow O_2 + \text{plant matter} \qquad (1.1)$$

The atoms, on the two sides of the above reaction, would add up stoichiometrically, provided the "plant matter" would contain 1 carbon atom, 2 hydrogen atoms and 1 oxygen atom (CH_2O), and nothing else. This is not really true; plants also contain other atoms, but we shall return to this later. To make the left and right side of the reaction balanced (Eq. 1.1) includng light (absorbed, to be precise), we should make additional

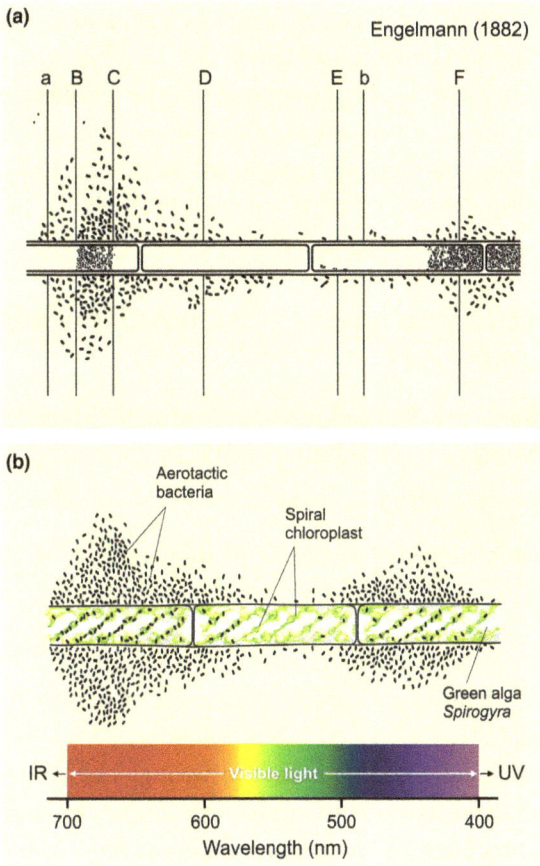

Fig. 1.1. Engelmann's experiments. **(a)** Demonstration of the first action spectrum of oxygenic photosynthesis, revealing Chl as the light-absorbing and photosynthesis-driving pigment. The above figure has been reproduced from the original drawing by Engelmann [1882]; it shows the filamentous green alga, *Cladophora* (two whole cells and parts of two other cells), and oxygen-seeking bacteria (depicted as small dots) that swim toward places where there is high O_2 concentration under the spectrum of visible light. The algal filament was placed parallel to the projected spectrum. The vertical lines with letters are positions of spectral lines that Engelmann used to calibrate his wavelength scale ($a = 718$ nm, $B = 687$ nm, $C = 656$ nm, $D = 589$ nm, $E = 527$ nm, $b = 518$ nm, $F = 486$ nm). The accumulation of aerotactic bacteria largely occurred in the red and in the blue regions of light. **(b)** Identification of cellular organelles, chloroplasts, as a site of photosynthetic oxygen production. In this experiment Engelmann used *Spirogyra*, another green alga, which has typical spiral chloroplasts, distinctly separated from other parts of the algal cells (note that *Cladophora*'s chloroplasts are uniformly spread in the cells). Again, the aerotactic bacteria moved to red and blue light regions of illuminated cells. However, this time, they were not randomly spread along these cells, but accumulated specifically in the chloroplast areas. IR = infrared light; UV = ultraviolet light.

9

adjustment, since we have a loss of energy as heat, and even some as light itself (fluorescence, to be discussed later).

We shall return later to the question about the amount and fate of the energy. For now, we shall concentrate on the atoms. We can see that on the left side of the formula there are oxygen atoms in both CO_2 and in water (H_2O); and on the right side there is not only molecular O_2 but also carbon and oxygen atoms in the plant matter.

1.2.3.3 Which atoms go where? Is it possible to answer this question?

At the beginning of the 20th century it was (wrongly) thought that the first step in photosynthesis was a splitting of CO_2 by light:

$$CO_2 + \text{light} \rightarrow O_2 + C^* \tag{1.2}$$

where, C^* denotes "activated carbon"; this was suggested to be followed by a second step:

$$n\,C^* + n\,H_2O \rightarrow C_nH_{2n}O_n \tag{1.3}$$

where $C_nH_{2n}O_n$ stands for a carbohydrate; with $n = 6$, we have a hexose, such as glucose, $C_6H_{12}O_6$.

The first indication that this is indeed wrong was obtained, not from plants, but from the discovery by Cornelis Bernardus van Niel (1897–1985) that some (sulfur) bacteria carry out a different kind of photosynthesis, without the evolution of oxygen, which can be summarized by the following simplified formula [Van Niel, 1941]:

$$n\,CO_2 + n\,H_2S + \text{light} \rightarrow 2n\,S + C_nH_{2n}O_n + n\,H_2O \tag{1.4}$$

Instead of evolving gaseous O_2, these sulfur bacteria deposit solid sulfur, an element closely related to oxygen, having the next place in the same column in the periodic table of elements. Here, water appears on the right side instead of the left side of the formula, i.e., as a product in the above reaction. Of course, the sulfur on the right side comes from hydrogen sulfide, H_2S.

If we assume that plants use H_2O in the same way as the bacteria use H_2S, we would write the formula as:

$$n\,CO_2 + n\,H_2O + \text{light} \rightarrow n\,O_2 + C_nH_{2n}O_n + n\,H_2O \tag{1.5}$$

And we would think that the oxygen evolved originated from water, just as the sulfur deposited by the bacteria came from hydrogen sulfide. This certainly is not a proof for the origin of oxygen in oxygenic photosynthesis; it is just the way scientists started to change their mind (see, e.g., [de Fourcroy, 1787; Wurmser, 1930; Joliot *et al.*, 2016]).

Robert (Robin) Hill [1937, 1939], of England, had prepared a suspension of ground plant leaves to which he added ferric oxalate, and, found that this sample produced oxygen when it was illuminated. Hill did not use the oxygen-loving bacteria that Engelmann had used (see Section 1.2.3.1; Fig. 1.1), but an equally ingenious analysis method, which in contrast to Engelmann's method was quantitative, although less sensitive. Hill added to his mixture a small amount of the red blood pigment, hemoglobin. When hemoglobin combines with oxygen, its absorption spectrum changes, and by measuring this change, Hill determined the amount of oxygen formed. Since it is known that ferric oxalate itself can undergo a photochemical reaction when exposed to light, Hill checked (and found) that there was oxygen production, when red light, which is not absorbed by ferric oxalate, but by Chl, was used. This "artificial" oxygen production has been called the *Hill reaction*. Here, instead of CO_2, ferric ion (Fe^{3+}) is reduced to ferrous ion (Fe^{2+}). The oxygen in this reaction cannot come from CO_2, which is not consumed, and thus, there is a strong indication that it comes from water, which is oxidized during the reaction:

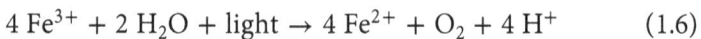

$$4\ Fe^{3+} + 2\ H_2O + light \rightarrow 4\ Fe^{2+} + O_2 + 4\ H^+ \qquad (1.6)$$

In the above reaction, four electrons are transferred from the two oxygen atoms in the two molecules of water to four iron ions.

There are different kinds of oxygen atoms, i.e., several isotopes of oxygen, and it was thought that one should be able to use "isotopically labeled" oxygen to prove conclusively that the oxygen released in photosynthesis is derived from water and not from CO_2 or bicarbonate ions (HCO_3^-), derived from CO_2. We shall later see how isotopically labeled carbon atoms were very important in elucidating the "Path of Carbon" in photosynthetic carbon assimilation. Of oxygen the most common natural isotope has the atomic weight of 16 (this is, in fact, what defines the atomic weight scale). Another stable, i.e., not radioactive, isotope has an atomic weight of 18, and it can be prepared at a concentration much higher than it is in natural water or oxygen. One of its first uses was when Ruben *et al.* [1941] attempted to solve the problem of the origin of oxygen released in photosynthesis. This proved to be not very straightforward. The main reason for this is the presence of carbonic anhydrase, an enzyme which catalyzes an exchange of oxygen atoms between water and CO_2 [Moroney *et al.*, 2001]. Another fact that may confuse the system is that the Hill reaction is stimulated by bicarbonate ions (see Section 3.3.2.3 in Chapter 3, and a review by Shevela *et al.* [2012]). But finally, with careful experimentation under suitable conditions, it has been shown that when bicarbonate-containing ^{18}O is administered, the evolved oxygen contains only ^{16}O [Stemler and Radmer, 1975; Clausen *et al.*, 2005; Hillier *et al.*, 2006]. Thus the O_2 is not derived from the bicarbonate.

We now return to the stoichiometric relation between the CO_2 consumed and the O_2 produced in oxygenic photosynthesis. The ratio between these quantities had been assumed to be 1. In many cases this is only a good approximation [Kaplan and Björkman, 1980], but in other cases large deviations can occur. One reason can be that the electrons derived from water are used for the assimilation of compounds other than the CO_2, and that the product of the reaction is something other than a carbohydrate. For instance, many plants reduce nitrate ions absorbed from the soil, while photosynthesis is going on in the leaves. In strong light, the ratio of O_2 evolved to CO_2 consumed can be as high as 1.26 [Bloom *et al.*, 1989]. Another reason for deviation from 1 can be in a system where the main products of photosynthesis are lipids rather than carbohydrates; here also the ratio would be above 1 since less oxygen, in relation to carbon, is bound in the lipids as compared to that in the carbohydrates. In many cases this ratio varies significantly, during the day. This is particularly obvious for the so-called Crassulacean Acid Metabolism (CAM) plants, which we shall discuss in Chapter 4.

Cyanobacteria (which were earlier called blue-green algae) carry out photosynthesis, i.e., oxygenic photosynthesis, as plants do [Shevela *et al.*, 2013]. But they have another ability that plants lack: they can photosynthetically assimilate not only CO_2, but also another atmospheric constituent, nitrogen, N_2. In some species, nitrogen is taken up by and assimilated by the same cells that take up CO_2, but in other species N_2 is taken up by special cells called heterocysts, which lack the ability to take up CO_2.

Thus far, we have dealt with the proportions only of the various atoms participating in photosynthesis, while we have said nothing about how much light is required, or how efficiently it is used. We shall return to this question in another context later, but for now we shall deal first with the light as we have dealt with matter, i.e., as particles.

1.2.3.4 How much light is required to drive photosynthesis?

Isaac Newton (1642–1727) had proposed in his famous 1704 book on "Opticks" that light consists of particles, but during the following years experimental evidence led scientists to regard it to have wave-motion (see a republished version: [Newton, 1998]). However, Einstein [1905] definitely showed that light also has particle properties, and since then physicists have agreed that light has both wave-like and particle-like properties, and that neither model completely depicts reality. Einstein was awarded the Nobel Prize in 1921 *"For his services to theoretical physics, and especially for his discovery of the law of the photoelectric effect,"* i.e., primarily for the paper quoted here. The light particles are called *photons*. Thus, our question boils down to: How many photons should we put in to replace the word *"light"* in our equation for photosynthesis?

For a long time there was disagreement on this (for historical papers, see [Nickelsen and Govindjee, 2011; Hill and Govindjee, 2014]), but finally it was established that under optimal conditions, the minimum number is "about 10" photons per oxygen evolved [Emerson and Lewis, 1941; Emerson and Chalmers, 1955; Emerson, 1958; Govindjee *et al.*, 1968]. We also now know this from the detailed knowledge of how the photochemical apparatus is constructed and how it functions; this will be described in later chapters. All the available data confirm that the minimum number of photons, needed to evolve one O_2 molecule, cannot be much lower than 10. Obviously, the number of incident photons is higher, since some light is reflected and some light is transmitted. Further, a small fraction of the absorbed light gives rise

13

to light emission (fluorescence) and some of the absorbed light is converted to heat [Govindjee and R. Govindjee, 1974].

In the next chapter we shall present our current view on the structure of the photosynthetic apparatus.

References

Blankenship, R. E. (2014). *Molecular Mechanisms of Photosynthesis*, 2nd ed. (Wiley-Blackwell, Hoboken, NJ).

Bloom, A. J., Caldwell, R. M., Finazzo, J., Warner, R. L. and Weissbart, J. (1989). Oxygen and carbon dioxide fluxes from barley shoots depend on nitrate assimilation, *Plant Physiol.*, 91, pp. 352–356.

Cahan, D. (2012). The awarding of the Copley Medal and the "discovery" of the law of conservation of energy: Joule, Mayer and Helmholtz revisited, *Notes Rec. R. Soc.*, 66, pp. 125–139.

Clausen, J., Beckmann, K., Junge, W. and Messinger, J. (2005). Evidence that bicarbonate is not the substrate in photosynthetic oxygen evolution, *Plant Physiol.*, 139, pp. 1444–1450.

de Fourcroy, M. (1787). *Bibliotheque universelles des Dames. Principes de Chimie*, Vol. II (Rue et Hotel Serpente, Paris).

de Saussure, N. T. (1804). *Recherches chimiques sur la vegetation* (Chez la Ve. Nyon, Paris).

Einstein, A. (1905). Über einen die Erzeugung und Verwandlung des Lichtes betreffenden heuristischen Gesichtspunkt, *Ann. Phys.*, 322, pp. 132–148. (Several English translations are available, including Arons, A. B. and Peppard, M. B. (1965) Einstein's proposal of the photon concept—A translation of the Annalen der Physik paper of 1905. *Am. J. Phys.*, 33, pp. 367–374.)

Emerson, R. (1958). The quantum yield of photosynthesis, *Annu. Rev. Plant Physiol.*, 9, pp. 1–24.

Emerson, R. and Chalmers, R. (1955). Transient changes in cellular gas exchange and the problem of maximum efficiency of photosynthesis, *Plant Physiol.*, 30, pp. 504–529.

Emerson, R. and Lewis, C. M. (1941). Carbon dioxide exchange and the measurement of the quantum yield of photosynthesis, *Am. J. Bot.*, 28, pp. 789–804.

Engelmann, T. W. (1882). Ueber Sauerstoffausscheidung von Pflanzenzellen im Mikrospektrum, *Bot. Z.*, 40, pp. 419–426.

Govindjee and Govindjee, R. (1974). Primary events in photosynthesis, *Sci. Am.*, 213, pp. 74–83.

Govindjee, R., Rabinowitch, E. and Govindjee. (1968). Maximum quantum yield and action spectrum of photosynthesis and fluorescence in chlorella, *Biochim. Biophys. Acta*, 162, pp. 539–544.

Helmholtz, H. (1847). *Ueber die Erhaltung der Kraft, eine physikalische Abhandlung* (G. Reimer Verlag, Berlin).

Hill, J. F. and Govindjee (2014). The controversy over the minimum quantum requirement for oxygen evolution, *Photosynth. Res.*, 122, pp. 97–112.

Hill, R. (1937). Oxygen evolved by isolated chloroplasts, *Nature*, 139, pp. 881–882.

Hill, R. (1939). Oxygen produced by isolated chloroplasts, *Proc. R. Soc. Lond. B*, 127, pp. 192–210.

Hillier, W., McConnell, I., Badger, M. R., Boussac, A., Klimov, V. V., Dismukes, G. C. and Wydrzynski, T. (2006). Quantitative assessment of intrinsic carbonic anhydrase activity and the capacity for bicarbonate oxidation in photosystem II, *Biochemistry*, 45, pp. 2094–2102.

Howe, H. M. (1965). A root of van Helmont's tree, *Isis*, 56, pp. 408–419.

Joliot, P., Crofts, A. R., Björn, L. O., Yerkes, C. T. and Govindjee. (2016). In photosynthesis, oxygen comes from water: From a 1787 book for women by Monsieur De Fourcroy, *Photosynth. Res.*, 129, pp. 105–107.

Kaplan, A. and Björkman, O. (1980). Ratio of CO_2 uptake to O_2 evolution during photosynthesis in higher plants, *Zeit. Pflanzenphysiol.*, 96, pp. 185–188.

Kipnis, N. (2014). Thermodynamics and mechanical equivalent of heat, *Sci. Educ.*, 23, pp. 2007–2044.

Krikorian, A. D. and Steward, C. (1968). Water and solutes in plant nutrition: With special reference to Van Helmont and Nicholas of Cusa. *Bioscience*, 18, pp. 286–292.

Lopes Coelho, R. (2009). On the concept of energy: How understanding its history can improve physics teaching, *Sci. Educ.*, 18, pp. 961–983.

Maxwell, J. C. (1865). A dynamical theory of the electromagnetic field, *Phil. Trans. R. Soc. Lond.*, 155, pp. 459–512.

Mayer, J. R. (1842). Bemerkungen über die Kräfte der unbelebten Natur, *Natur. Justus Liebigs Ann. Chem.*, 42, pp. 233–240.

Mayer, J. R. (1845). *Die organische Bewegung in ihrem Zusammenhange mit dem Stoffwechsel. ein Beitrag zur Naturkunde* (Drechsler, Heilbronn).

Moroney, J. V., Bartlett, S. G. and Samuelsson, G. (2001). Carbonic anhydrases in plants and algae, *Plant Cell Environ.*, 24, pp. 141–153.

Newton, I. (1998). *Opticks: or, a treatise of the reflexions, refractions, inflexions and colours of light. Also two treatises of the species and magnitude of curvilinear figures. (Contains a digitized facsimile of the edition of 1704, reproduced from the copy in the Warnock Library; with commentary by Nicholas Humez).* (Octavo Corporation, Palo Alto).

Nickelsen, K. (2015). *Explaining Photosynthesis: Models of Biochemical Mechanisms, 1840–1960*, Vol. 8 (Springer, Dordrecht).

Nickelsen, K. and Govindjee. (2011). *The Maximum Quantum Yield Controversy: Otto Warburg and the "Midwest-Gang"* (Bern Studies in the History and Philosophy of Science, Bern).

Rabinowitch, E. and Govindjee. (1969). *Photosynthesis* (John Wiley & Sons Inc., New York).

Ruben, S., Randall, M., Kamen, M. and Hyde, J. L. (1941). Heavy oxygen (O^{18}) as a tracer in the study of photosynthesis, *J. Am. Chem. Soc.*, 63, pp. 877–879.

Severinghaus, J. W. (2014). Crediting six discoverers of oxygen. *In* Swartz, H. M., Harrison, D. K., Bruley, D. F., eds, *Oxygen Transport to Tissue XXXVI* (Springer, New York, NY), pp. 9–17.

Shevela, D., Eaton-Rye, J. J., Shen, J.-R. and Govindjee. (2012). Photosystem II and the unique role of bicarbonate: A historical perspective, *Biochim. Biophys. Acta*, 1817, pp. 1134–1151.

Shevela, D., Pishchalnikov, R. Y., Eichacker, L. A. and Govindjee. (2013). Oxygenic photosynthesis in cyanobacteria. *In* Srivastava, A. K., Amar, N. R., Neilan, B. A., eds, *Stress Biology of Cyanobacteria: Molecular Mechanisms to Cellular Responses* (CRC Press/Taylor & Francis Group, Boca Raton, FL), pp. 3–40.

Stemler, A. and Radmer, R. (1975). Source of photosynthetic oxygen in bicarbonate-stimulated Hill reaction, *Science*, 190, pp. 457–458.

van Helmont, J. B. (1648). *Ortus medicinae* (Apud Ludovicum Elzevirium, Amsterdam).

van Niel, C. B. (1941). The bacterial photosyntheses and their importance for the general problem of photosynthesis. *In* Nord, F. F., Werkman, C. H., eds, *Advances in Enzymology and Related Areas of Molecular Biology*, Vol. 1 (John Wiley & Sons, Inc., Hoboken, NJ), pp. 263–328.

Wurmser, R. (1930). *Oxydations et réductions* (Les Presses Universitaires de France, Paris).

Chapter 2

The Photosynthetic Apparatus

2.1 Introduction

Photosynthesis is a complex process and cannot, and does not, occur in a homogeneous solution, but requires a complex structural matrix; for one, a structure is needed to direct reactions in the right way. In a light microscope, we only have a resolution of ~200 nm, but under an electron microscope it is thousand-fold higher, ~0.2 nm. An understanding of photosynthesis requires a description of the structural and physico-chemical details of various photosynthetic membranes as well as their components. We shall start with chloroplasts.

2.2 Chloroplasts

Using an ordinary light microscope, we can see that there are, inside the cells of algae and plant leaves, green organelles, *chloroplasts*; their shape is quite diverse among algae. In higher plants, a chloroplast looks like a lens (Fig. 2.1). Chloroplasts contain, among other components, all the *chlorophyll* molecules. Further, it is where all the reactions of oxygenic photosynthesis take place.

To examine chloroplasts in their details, we use an electron microscope, which has a much higher resolution than a light microscope, since the wavelength corresponding to the electrons used are a thousand times shorter (~0.2 nm) than those of light (~500 nm; cf. Section 2.1). Fig. 2.2a

Photosynthesis: Solar Energy for Life by Dmitry Shevela, Lars Olof Björn and Govindjee
© 2018, published by World Scientific Publishing Co. Pte. Ltd. ISBN: 978-981-3223-10-3.

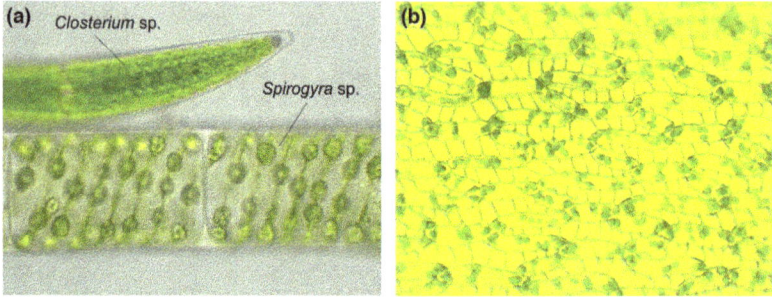

Fig. 2.1. Light micrographs of chloroplasts from three different organisms. **(a)** Filamentous green alga *Spirogyra* sp. with its typical helically arranged chloroplasts (only two cells of the filament are shown) and, on top of it, unicellular green alga *Closterium* sp. with two characteristic chloroplasts, separated in the middle of the cell by the nucleus. **(b)** Cells of moss *Sphagnum* sp., with several spherical shaped chloroplasts in each cell. Courtesy of Natalia Voronkina; reproduced with permission.

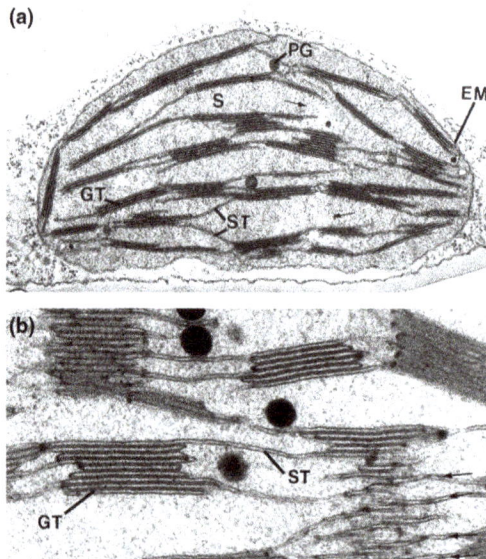

Fig. 2.2. Electron micrographs of chloroplasts from tobacco (a) and spinach (b), at two different magnifications. **(a)** Thin section electron micrograph of a young, developing tobacco chloroplast. Two envelope membranes (EM) delineate the chloroplast stroma (S) within which stacked grana thylakoids (GT) and unstacked stroma thylakoids (ST) can be recognized. Plastoglobuli (PG) and DNA containing regions (small arrows) are also seen. **(b)** Thin section through a portion of a spinach chloroplast showing interconnected GT and nonstacked ST membranes. Note the flattened, sac-like structure of the individual thylakoids. From Staehelin and van der Staay [1996]. Courtesy of Andrew Staehelin; reproduced with permission.

shows what a longitudinal cut through a tobacco (*Nicotiana* sp.) chloroplast looks like under an electron microscope, whereas Fig. 2.2b shows details, at a higher magnification, of a spinach (*Spinacia* sp.) chloroplast. Chloroplasts are surrounded by a double membrane, inside of which there is an elaborate membrane system, referred to as *thylakoids* (from Greek Θυλακοιδ). Originally it was thought that each thylakoid was a separate closed sac (a pouch) or a vesicle; later it was realized that the thylakoids form a connected membrane system (Fig. 2.3). Each thylakoid separates an internal space called *lumen*, from an outer space, a matrix called *stroma*, which has, among other components, the necessary enzymes for carbon fixation. Further, there is a stack of thylakoids, called a *granum*. Here and there the thylakoid membranes lie stacked together to form *grana*; at other places, pairs of them are separated by stroma.

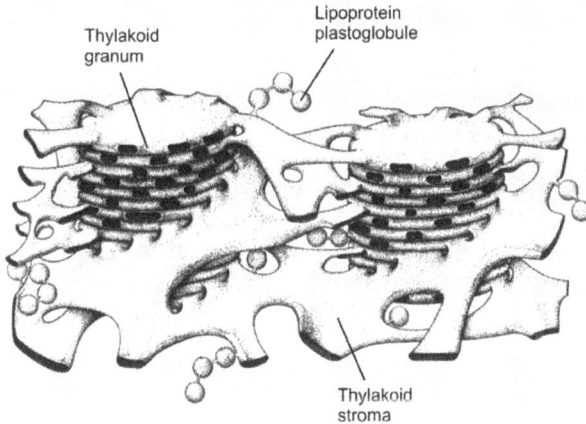

Fig. 2.3. Three-dimensional model of the spatial relationship between grana and stroma thylakoids. The drawing shows two grana stacks interconnected by parallel sets of non-stacked stroma thylakoids that spiral up and around the stacks in a right-handed helical conformation. Where the stroma thylakoids intersect with the grana thylakoids, the two types of membrane domains are connected through tubular junctions that vary in length. The lipoprotein plastoglobule particles are attached to the thylakoids and to each other through a half-lipid bilayer that surrounds the content of the globule, and is continuous with the stroma-side leaflet of the thylakoid membrane. This unpublished model of L. A. Staehelin is based on Austin *et al.* [2006] and Austin and Staehelin [2011]; reproduced with permission.

2.3 Thylakoid Membrane and Photosynthetic Protein Complexes

Thylakoid membranes are double layers of polar lipid, with their lipophilic "fat-loving" ends turned toward one another, and the hydrophilic "water-loving" end(s) toward the stroma (or, in the grana, toward the hydrophilic side of another membrane). In these lipid membranes, several kinds of large protein complexes are embedded. In contrast to most other membranes in bacteria and eukaryotes, which contain a large amount of phospholipids, thylakoid membranes contain galactolipid(s) as the main lipid component outside the protein complexes [Lichtenthaler and Park, 1963; Quinn and Williams, 1983]. When the proteins change positions, they do it in a highly regulated fashion, as we shall discuss later.

In Fig. 2.4, we show a cartoon of the protein complexes in a thylakoid membrane. Most of the chlorophylls, the green pigments, and the *carotenoids*, the yellow-to-orange pigments, are part of both the *photosystem I* (PSI) and the *photosystem II* (PSII) complexes. Two other protein complexes on the thylakoids are: a complex called the *cytochrome b_6f complex* (Cyt b_6f), equivalent to Cyt *bc* complex of *mitochondria*, and the *ATP synthase* (ATPase). All the four complexes, mentioned above, are located in the thylakoid membrane, and occupy the entire space from the stroma side to the lumen side (for further details, see Chapter 3).

The thylakoid membrane of all oxygenic organisms is the site of the photosynthetic "light reactions" (see Fig. 2.5 and for more details, see Chapter 3). The use of "light reactions" does not imply that all the reactions require light; in fact, the only *real* light reaction is the absorption of light by chlorophyll molecule and the formation of its excited singlet state; one could say that the "light reaction" is over once light energy has been converted to chemical energy at the reaction centers. In algae and plants, the thylakoid membrane is located in chloroplasts, but in *cyanobacteria* it is present within the cytoplasm. The end result of the light reactions is the production of NADPH and ATP; these two drive the so-called dark reactions (black arrows in Fig. 2.5) of CO_2 fixation in stroma of the chloroplast *via* a metabolic pathway, known as the Calvin–Benson–Bassham cycle, or

20

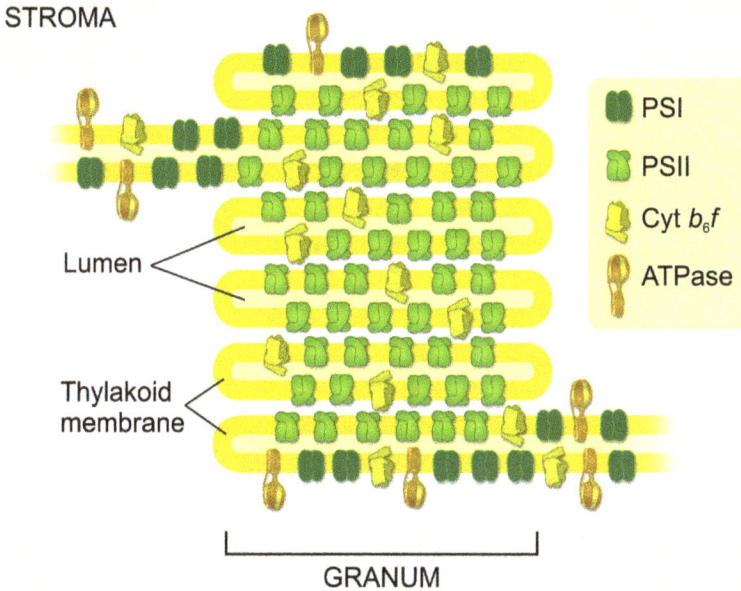

Fig. 2.4. An illustration showing the location of various photosynthetic protein complexes in grana-thylakoids and in stroma-thylakoids (between the grana). Abbreviations: ATPase, ATP synthase; Cyt b_6f, cytochrome b_6f; PSI, photosystem I; PSII, photosystem II. For simplicity, all PSII complexes are shown as monomers. Also, light harvesting complexes are not included in this model. Based on Dekker and Boekema [2005] and Pribil *et al.* [2014].

the Calvin–Benson cycle (called by some as the Calvin cycle , which we do not approve since it leaves out the major player Benson). This results in the reduction of CO_2 to energy-rich carbohydrates (e.g., sucrose and starch). Chapter 4 provides a detailed overview of these "dark reactions." Note that the term "dark reactions" implies only that these reactions are not *directly* driven by light; they also take place in light: in fact, light is needed to activate several enzymes. However, see Buchanan [2016] for a perspective, where the use of "dark reactions" is challenged; instead "carbon reactions" is preferred.

In Chapter 3, we shall describe in detail the protein complexes of the photosynthetic apparatus. We note that chloroplasts contain a large number of other proteins in addition to the four major membrane protein complexes (PSII; PSI; Cyt b_6f; and ATPase), as mentioned above

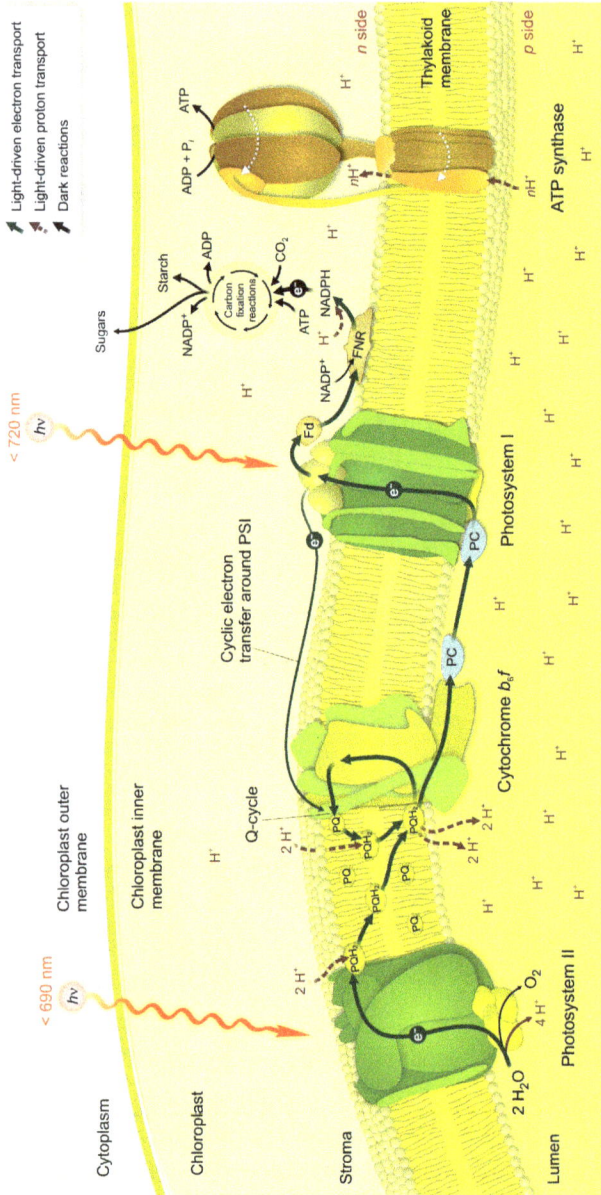

Fig. 2.5. A simplified schematic representation of a thylakoid membrane, photosynthetic protein complexes and some intermediates (metabolites) involved in the electron and the proton transport reactions of photosynthesis in higher plants and green algae. The dark green arrows indicate the light-driven electron transport, brown arrows indicate the proton flow, and some dark reactions are colored in black. *Note that the overall scheme is for the release of one oxygen molecule, i.e., transfer of 4 electrons from water to $NADP^+$, as well as the operation of the Q-cycle. Thus, each Photosystem (I and II) will need to receive at least 4 photons each.* Abbreviations: ADP, adenosine diphosphate; ATP, adenosine triphosphate; FNR, ferredoxin-$NADP^+$ oxidoreductase; $NADP^+$ and NADH, oxidized and reduced forms of nicotinamide-adenosine dinucleotide phosphate; P_i, orthophosphate; PQ, plastoquinone.

(see Fig. 2.5). These include a small, mobile, copper-containing protein in the lumen, plastocyanin (PC), serving as an electron carrier between the Cyt b_6f complex and the PSI (see Chapter 3). In the stroma, there are two other proteins: ferredoxin (Fd), which has an iron-sulfur center and ferredoxin:NADP oxireductase (FNR), which is a flavin enzyme (Fig. 2.5). We do not list here the names of the many enzymes, present in the stroma, and needed for the assimilation of carbon dioxide; these will be described in Chapter 4.

2.4 Pigments

Thylakoid membranes of higher (land) plants and green algae contain two major kinds of pigments: the chlorophylls (Chls) and the carotenoids (Cars), as already mentioned above. (We shall discuss later other pigments, present in phycobiliproteins of cyanobacteria, and algae.) In plants, there are only two kinds of chlorophylls, chlorophyll *a* (Chl *a*) and chlorophyll *b* (Chl *b*) (Fig. 2.6a), while some algae and bacteria contain other Chls (Chl *c*, Chl *d*, Chl *e*, and Chl *f*). In plants, the following Cars are usually present: β-carotene (largest amount), zeaxanthin, violaxanthin, lycopene (Fig. 2.6b), and small amounts of neurosporene and ζ-carotene, an acyclic isoprenoid with 11 double bonds (not shown).

2.5 Lipids and Proteins

The four main lipids found in thylakoid membranes are *monogalacto-syldiacylglycerol* (MGDG), *digalactosyldiacylglycerol* (DGDG), *sulfoqui-novosyldiacylglycerol* (SQDG), and *phosphatidylglycerol* (PG) (Fig. 2.7); these are quite different from those of, e.g., cell membranes in our own bodies, but they also have a hydrophilic and a lipophilic end. The total number of their molecules, in the thylakoid membrane, is almost twice that of the number of Chl molecules [Kirchhoff *et al.*, 2002]. A minor but very important lipid component in the thylakoid membranes is *plas-toquinone* (see Figs. 2.5 and 2.7, and Chapter 3). It serves as a carrier of electrons (from PSII to Cyt b_6f) and protons. The Chls and Cars are also lipids, but are almost always bound to the proteins.

(a)

Chlorophyll *a*

Chlorophyll *b*

Chlorin head (polar)

Hydrocarbon tail (non-polar)

(b)

β–carotene

Zeaxanthin

Violaxanthin

Lycopene

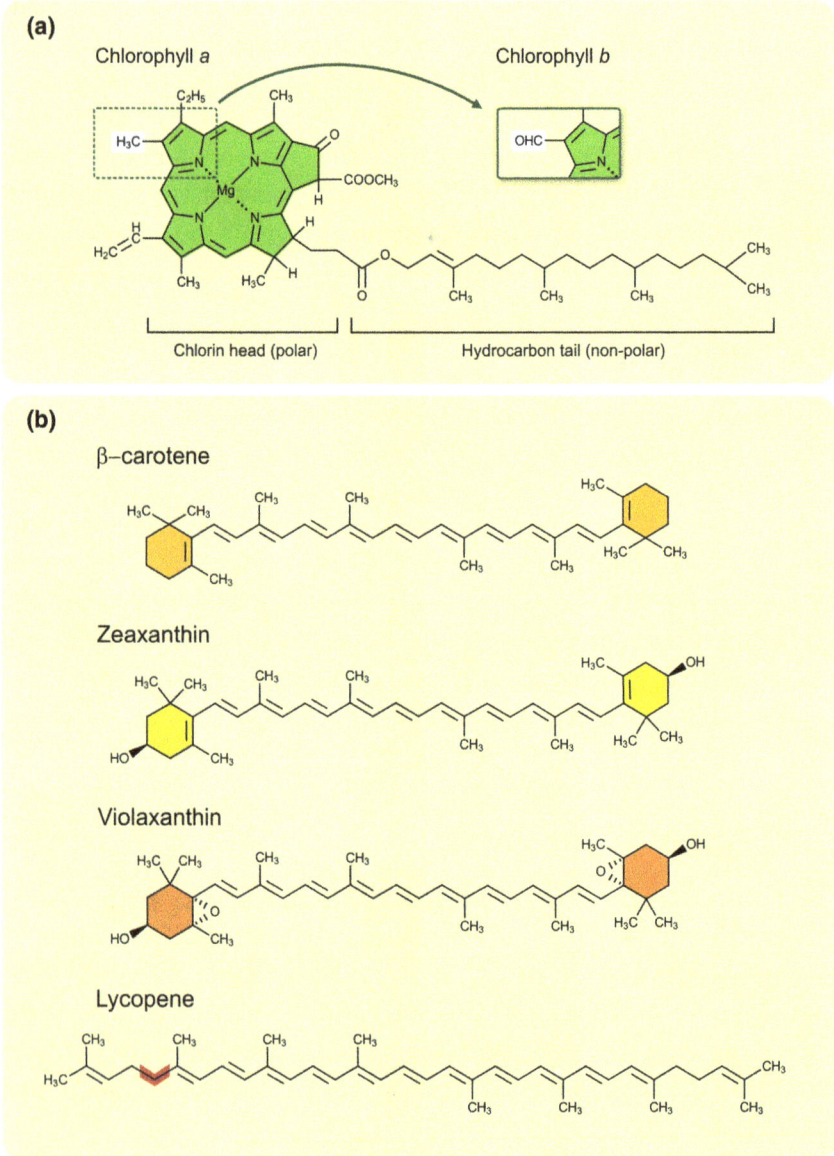

Fig. 2.6. Molecular structures of photosynthetic pigments. **(a)** Chlorophylls *a* and *b*. **(b)** Four different carotenoids.

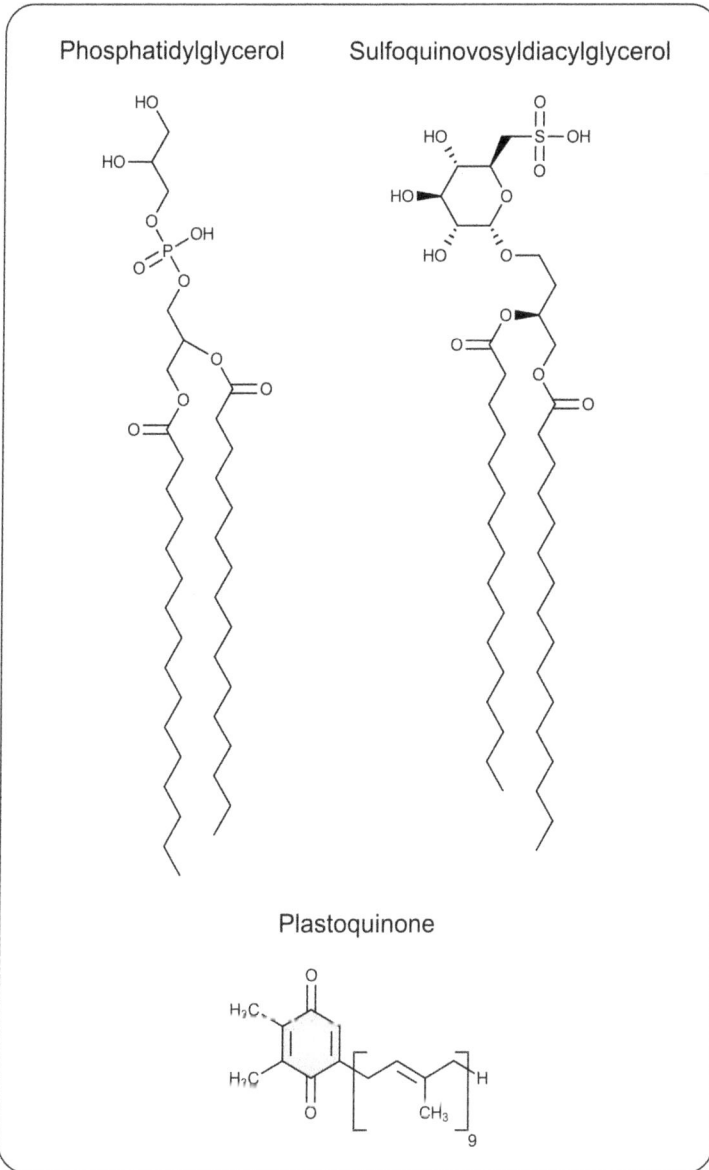

Fig. 2.7. Molecular structures of three lipids found in thylakoids.

2.6 Evolution

Plants have arisen during evolution by a "fusion" of three kinds of microorganisms, probably an *archeon*, a proteobacterium, and a cyanobacterium [Björn and Govindjee, 2015; Soo *et al.*, 2017]. The proteobacteria must have evolved to become mitochondria (the site of respiration), and the cyanobacteria must have evolved to become chloroplasts (and other plastids present in the roots, e.g., starch-containing amyloplasts, and the fruits, e.g., carotenoid-containing chromoplasts); in addition, many genes of these microorganisms must have been transferred to the cell nucleus. In addition to Chls and Cars, cyanobacteria use *phycobiliproteins* to absorb orange-green light for their photosynthesis. These have been retained in parts of the "red line" of evolution, i.e., in the red algae, and some other algae (e.g., cryptophytes), which have originated by fusion of red algae with other organisms. But they were lost on the "green line" of evolution, i.e., in green algae and land plants (see Chapter 8 and Figs. 8.1 and 8.2). For further discussion on evolution of photosynthesis, see Blankenship [2010], Hohmann-Marriott and Blankenship [2011], and Fischer *et al.* [2016].

In Chapter 3 we shall talk about light-dependent reactions which take place in all photosynthetic organisms.

References

Austin, J. R., Frost, E., Vidi, P.-A., Kessler, F. and Staehelin, L. A. (2006). Plastoglobules are lipoprotein subcompartments of the chloroplast that are permanently coupled to thylakoid membranes and contain biosynthetic enzymes, *Plant Cell*, 18, pp. 1693–1703.

Austin, J. R. and Staehelin, L. A. (2011). Three-dimensional architecture of grana and stroma thylakoids of higher plants as determined by electron tomography, *Plant Physiol.*, 155, pp. 1601–1611.

Björn, L. O. and Govindjee (2015). The evolution of photosynthesis and its environmental impact. *In* Björn, L. O., ed, *Photobiology: The Science of Light and Life* (Springer, New York), pp. 207–230.

Blankenship, R. E. (2010). Early evolution of photosynthesis, *Plant Physiol.*, 154, pp. 434–438.

Buchanan, B. B. (2016). The carbon (formerly dark) reactions of photosynthesis, *Photosynth. Res.*, 128, pp. 215–217.

Dekker, J. P. and Boekema, E. J. (2005). Supramolecular organization of thylakoid membrane proteins in green plants, *Biochim. Biophys. Acta*, 1706, pp. 12–39.

Fischer, W. W., Hemp, J. and Johnson, J. E. (2016). Evolution of oxygenic photosynthesis, *Annu. Rev. Earth Planet. Sci.*, 44, pp. 647–683.

Hohmann-Marriott, M. F. and Blankenship, R. E. (2011). Evolution of photosynthesis, *Annu. Rev. Plant Biol.*, 62, pp. 515–548.

Kirchhoff, H., Mukherjee, U. and Galla, H. J. (2002). Molecular architecture of the thylakoid membrane: Lipid diffusion space for plastoquinone, *Biochemistry*, 41, pp. 4872–4882.

Lichtenthaler, H. K. and Park, R. B. (1963). Chemical composition of chloroplast lamellae from spinach, *Nature*, 198, pp. 1070–1072.

Pribil, M., Labs, M. and Leister, D. (2014). Structure and dynamics of thylakoids in land plants, *J. Exp. Bot.*, 65, pp. 1955–1972.

Quinn, P. J. and Williams, W. P. (1983). The structural role of lipids in photosynthetic membranes, *Biochim. Biophys. Acta*, 737, pp. 223–266.

Soo, R. M., Hemp, J., Parks, D. H., Fischer, W. W. and Hugenholtz, P. (2017). On the origins of oxygenic photosynthesis and aerobic respiration in cyanobacteria, *Science*, 355, pp. 1436–1440.

Staehelin, L. A. and van der Staay, G. W. M. (1996). Structure, composition, functional organization and dynamic properties of thylakoid membranes. *In* Ort, D. R., Yocum, C. F., eds, *Oxygenic Photosynthesis: The Light Reactions* (Kluwer, Dordrecht), pp. 11–30.

Chapter 3

Basics of Photosynthesis: Light-Dependent Reactions

3.1 Overview: Harvesting Sunlight to Drive Redox Chemistry

In all photosynthetic organisms (both oxygenic and anoxygenic) the photo-synthetic light reactions begin with the absorption of light (photons) by pigments in light-harvesting complexes, embedded in the thylakoid membrane (or in case of cyanobacteria, also in phycobilisomes). The antenna (both outer and inner) systems deliver the energy of absorbed light (excitation energy; we shall discuss it in Section 3.2) to pigment–protein reaction center complexes, Photosystems II and I (PSII and PSI; see Section 3.3), both embedded in the thylakoid membrane (Fig. 3.1). As a consequence of primary photo-chemistry, which takes place after trapping of the excitation energy by special photoactive chlorophyll (Chl) molecules in the reaction centers of the two photosystems, light energy is converted into chemical energy. This energy drives the redox chemistry of the stepwise linear electron "transfer" from water to the oxidized form of nicotinamide adenine dinucleotide phosphate (NADP$^+$), involving PSII and PSI as well as the Cyt b_6f complex (Fig. 3.1). In this chapter we shall also briefly describe photosynthetic ATP production by ATP synthase from ADP and inorganic phosphate (see Section 3.3.3).

3.2 Capturing the Energy of Light

The first step in photosynthesis is the absorption of light by pigment molecules which include Chl a, other Chls, or phycobilins, or fucoxanthol,

Photosynthesis: Solar Energy for Life by Dmitry Shevela, Lars Olof Björn and Govindjee
© 2018, published by World Scientific Publishing Co. Pte. Ltd. ISBN: 978-981-3223-10-3.

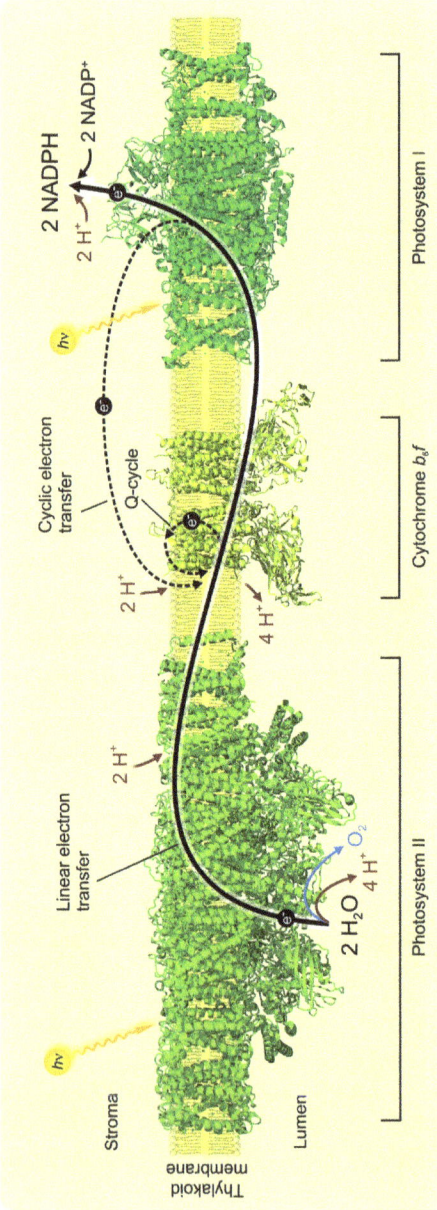

Fig. 3.1. A side-on view of the three major photosynthetic complexes (Photosystem II, PSII; Cytochrome b_6f, Cyt b_6f; and Photosystem I, PSI) embedded in the thylakoid membrane. These complexes bind the redox components (not shown) required for linear electron transfer (indicated by the big bold black arrow) from water to nicotinamide pyridine nucleotide phosphate, $NADP^+$. However, under some conditions, electrons on the electron acceptor side of PSI cycle back toward the Cyt b_6f complex rather than going to $NADP^+$ and then again back to PSI, thus-performing the *cyclic electron transfer* (dashed arrow; see Bendall and Manasse, 1995; Joliot *et al.*, 2006). As a result, no NADPH is produced, but the available energy is used for ATP synthesis. This is also the case for the so-called Q-cycle (dashed arrows) [Mulkidjanian, 2010; Cramer *et al.*, 2011] which increases the number of protons pumped across the membrane, per electron transferred (Fig. 2.5). The proton transfer steps are indicated by the brown arrows. The higher plant (here, spinach) dimeric PSII-LHCII supercomplex was visualized using the coordinates of cryo-EM structure at 3.2 Å deposited at the Protein Data Bank (PDB) with ID 3JCU [Wei *et al.*, 2016]. The Cyt b_6f complex (dimer) of the thermophilic cyanobacterium *Mastigocladus laminosus* was generated from a 3.0 Å crystal structure using PDB ID 1VF5 [Kurisu *et al.*, 2003]. The PSI-LHCI supercomplex from spinach at 2.8 Å resolution was produced employing coordinates from PDB, using ID 4Y28 [Mazor *et al.*, 2015]. See the legend of Fig. 2.5 about the stoichiometry of electrons transferred, and photons needed to run it.

depending on the organism; this step occurs within femtoseconds (one femtosecond is 10^{-15} s; there are as many femtoseconds in a second as there are seconds in 31.54 million years!). This event means that a photon disappears, and the energy of the molecule increases, the pigment molecule is in an excited state. We need to consider only two kinds of energy of the molecule here: the electronic energy and the vibrational energy, and both are changed when the photon is absorbed, as a consequence of the so-called *Franck-Condon principle*: Upon absorption of a photon, the electronic transition from the ground state to the excited state occurs without a change in the position of the nuclei, because "the electrons are light and the nuclei are heavy," and the molecule goes to a higher vibrational state (see Fig. 3.2, and the discussion that follows; for

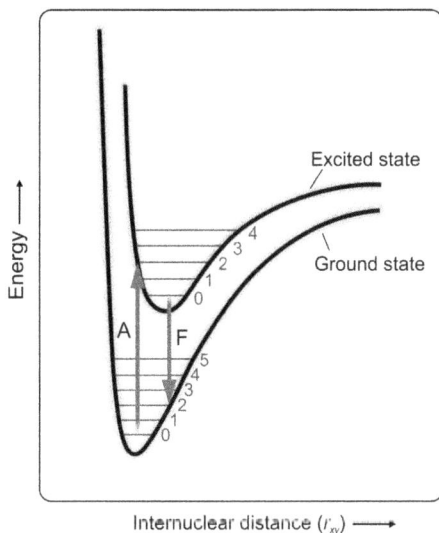

Fig. 3.2. Energy levels in a molecule. For simplicity we use a molecule with two atoms. The distance between the two atomic nuclei is plotted along the horizontal direction, and the energy in the vertical direction. The two thick black curves depict the potential energy in two different electronic states: ground and excited. Let us consider first the lower curve representing the ground state. When the atomic nuclei approach one another the potential energy rises because the nuclei repel one another. When they get far from one another, the energy also rises, because the electronic "rubber bands" stretch. The minimum potential energy, the lowest part of the curve, is at a point in between. Where the potential energy is lowest, the kinetic energy is highest. After excitation (A for absorption), the molecule goes very quickly to its lowest vibrational state. From here, we can have photochemistry or loose energy as fluorescence (F) or as heat.

Franck-Condon principle, see Rabinowitch and Govindjee [1969], and Atkins and Friedman [1999]). In addition, we have "vibronic" energy, which is a combination of "vibrational" and "electronic" energy: the "vibronic energy coupling" refers to the interaction between electronic and nuclear vibrational energy. This interaction, indeed, makes the light harvesting more efficient in several cases [Dean *et al.*, 2016].

The atomic nuclei in the molecule repel each other, because they are all positively charged, but they are held together by the negatively charged electrons surrounding them. We can think of the electron clouds between the nuclei as elastic rubber bands. The distance between two nuclei is not constant, but oscillating between a minimum and a maximum distance, and this oscillation provides them with vibrational energy. The vibrational energy can be regarded as an ongoing change between potential (positional) energy, and kinetic energy (energy of movement). The energy can assume only discrete values, i.e., it is quantized.

There are several horizontal lines, symbolizing various possible levels of vibrational energy (Fig. 3.2). The vibrational energy cannot assume values in between, because of the quantization of energy. When a photon is absorbed by a molecule in the lowest electronic (ground) state (the horizontal line associated with the lowest black curve in Fig. 3.2) the molecule goes to a higher (excited) electronic energy state (a change shown by the vertical arrow **A** to a horizontal line associated with the upper black curve). While still in the ground electronic state, the nuclei spend much of their time near the position where they have the lowest potential energy and the highest kinetic energy, i.e., at the place corresponding to the lowest dip of the curve. When a photon is absorbed, the electrons change their configuration so quickly that the nuclei are still at the same position immediately after the change. This is because nuclei are "heavy" and electrons are "light," as stated above. But in the new configuration this is not the position where the potential energy is lowest. The nuclei have acquired additional vibrational energy, but "fall" back (from 2 to 0 in Fig. 3.2) to the lowest vibrational level by transferring energy as heat, to neighboring nuclei. If the molecule now rids itself also of electronic energy (arrow **F** in Fig. 3.2) it will again arrive at a vibrational level that is not the lowest, so further energy will be lost as heat. Part of the electronic energy lost by the molecule, in the above process, appears as light (*fluorescence*). If so, the emitted photon will have lower energy corresponding to increased wavelength than the originally absorbed one (undergoing a "red shift"). The energy difference between absorbed and emitted photon corresponds to the difference in length between arrows **A** and **F**. This is the so-called *Stokes shift*, named after Sir George G. Stokes.

In photosynthetic organisms, some of the absorbed energy does, in fact, appear as fluorescence, but most of the energy is converted to chemical energy. First, we must say a little more about the Chl molecule, which is much more complicated than the simple molecule consisting of two atoms that was used in Fig. 3.2. Instead of two, a Chl molecule consists of 137 (Chl *a*) or 136 (Chl *b*) atoms. We can easily understand that a complete energy diagram for a Chl molecule would be much more complicated than that shown in Fig. 3.2, but we shall still use another simplified diagram (see Fig. 3.3).

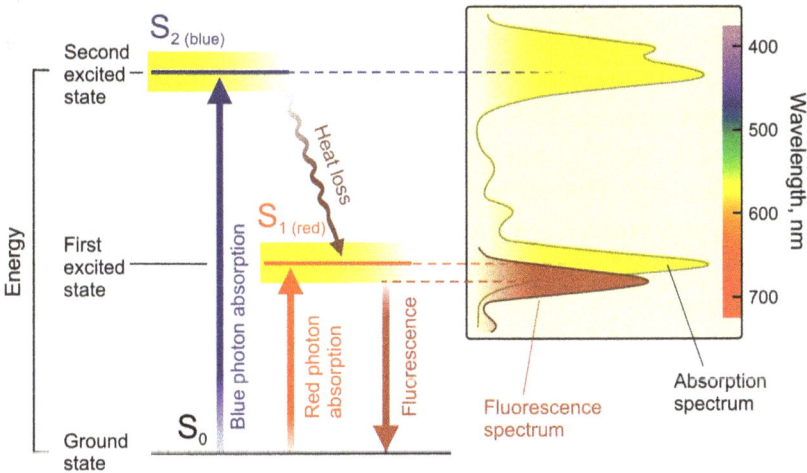

Fig. 3.3. A simplified "Jablonski-Perrin*" diagram of the energy levels of a Chl molecule with spectral transitions between them (vertical arrows) and absorption/fluorescence spectrum (turned 90° from the usual orientation) of Chl *a* corresponding to these levels. Note that, for the sake of simplicity, intersystem crossing to the triplet state of Chl, as well as vibrational sublevels of the electronic energy levels of Chl are not shown. [*Named after Aleksander Jabłoński of Poland and Francis Perrin of France.]

The absorption spectrum of Chl, shown on the right in Fig. 3.3, has two main bands, basically with two peaks, corresponding to two major energy levels (S_1 and S_2) above the ground level in the Chl molecule, as shown on the left in the figure. [In reality, there are other energy levels between S_1 and S_2, and S_2 is listed as S_n, but we will keep it simple here.] The highest of these, called the second excited state (S_2), corresponds to the absorption in the blue part of the spectrum. The lower level, the first excited state (S_1), corresponds to the absorption band in the red part of the spectrum. [We remind ourselves of the inverse relation between "E," the energy, and the wavelength of light: $E = h \times c/\lambda$, where "h" is Planck's constant, "c" the velocity of light, and "λ" the wavelength.]

The second excited state of Chl, which is reached by absorption of blue light, is very short-lived, and decays in about 10^{-13} s [Reimers *et al.*, 2013] to the first excited state by conversion of some of the energy to vibrational energy. The first excited state can also be reached in this manner, but mostly by absorption of red light (Fig. 3.3).

We note that only the first excited state of Chl *a* directly provides most of the energy for photosynthesis, some for fluorescence, and some for heat. A considerable fraction of the energy of blue light absorbed by the plant is lost as heat (Fig. 3.3). Although Chl *a* weakly absorbs green light between the blue and red parts of the spectrum, this absorption is also important for the plant. Red and blue light are absorbed near the upper surface of a leaf, while green light penetrates deeper and provides energy to chloroplasts in the interior of the leaf and near the lower surface.

Most of the Chl *a* molecules (or, in some cases, other molecules) that have absorbed light do not take direct part in any photochemical reaction. They act as *antenna molecules*, and transfer their absorbed energy to other pigment molecules, but eventually the energy ends up in *reaction center Chl a molecules*, where the photochemical reaction takes place (see Section 3.3). This transfer of excitation energy from one pigment molecule to another can take place by various mechanisms. The most important mechanism in plants is called the *Förster mechanism* or *resonance transfer (often referred to as FRET, Förster resonance energy transfer)*, involving dipole–dipole interaction, while other mechanisms dominate in some photosynthetic bacteria.

We can understand the Förster mechanism in the following way: Transition of a molecule from one electronic energy state to another, a transition during which some electrons change their orbitals, causes a change in the electrical field around the molecule. Conversely, a change in the electric field around a molecule can cause its transition from one energy state to another. The field change caused by the transition in one molecule can cause the opposite transition in a neighboring molecule.

The intensity of the change of the electrical field around a molecule, when it changes its energy state, drops off with the third power of the distance (r^{-3}) from the molecule. Also the sensitivity of a molecule to "experience," or to "feel" a field change drops off with the third power of the distance (r^{-3}). Thus, the effect that an electronic change in one molecule has on the electronic state in a neighboring molecule drops off with the sixth power of the distance (r^{-6}) in the Förster mechanism; this then governs *excitation energy transfer* from a *donor* molecule to an *acceptor* molecule; this transfer also depends on the "overlap integral," a measure of how similar the possible energy jumps in the molecules are. Furthermore, the orientation of the molecules in relation to the others is also very important here. In summary, the transfer rate is proportional to the following expression:

$$(\text{fluorescence quantum yield}) \times (\text{overlap integral}) \times \kappa^2 \times r^{-6} \qquad (3.1)$$

where r is the distance between the donor and the acceptor molecules, and κ^2 is a geometric factor which depends on the angle between the transition moments involved in the energy transfer, and also on the angles between them and the line connecting the dipole centers of the molecules (see, e.g., Fig. 5 in Mirkovic *et al.*, 2017). The fluorescence quantum yield is the fraction of excitation quanta that would appear as fluorescence photons from the molecule giving off energy (the *donor*) in the absence of a molecule taking up the energy (the *acceptor*) (Fig. 3.4). Its inclusion in the formula accounts for energy being lost as heat. As implied above, the overlap integral is a measure of how much the fluorescence spectrum of the donor overlaps with the absorption spectrum of the acceptor. For a complete discussion of light absorption and excitation energy transfer in photosynthesis, see Mirkovic *et al.* [2017].

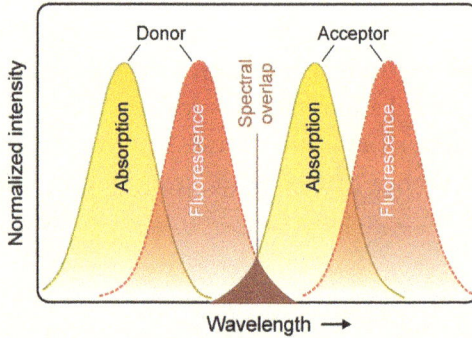

Fig. 3.4. Schematic representation of the spectral overlap between the fluorescence spectrum of a donor molecule and the absorption spectrum of an acceptor molecule. Note that the area marked "Spectral overlap" (common energy levels shared between the donor and acceptor) is not quite proportional to the so-called "overlap integral."

3.3 Conversion of Light Energy into Chemical Energy

3.3.1 *Primary photochemistry*

Finally the energy that has been hopping around (in some cases, in a coherent manner) among antenna molecules arrives at a special pair of Chl *a* molecules called a *reaction center* (RC). In plants, as well as algae and cyanobacteria, there are two kinds of RCs, reaction center I (RCI) and reaction center II (RCII). They are located in the large protein complexes mentioned in Chapter 2, PSI and PSII, respectively. In Fig. 3.5 we show the structures of PSI and PSII, at this stage mainly to give the reader an impression of the enormous complexity of these supercomplexes, and to show the Chl molecules, embedded in proteins. We shall return to some of these components later in this book.

In the RC Chl *a* molecules, the energy needed to raise an electron to its higher energy level is lower than in other pigment molecules in the antenna. At these Chl *a* molecules, a *charge separation* takes place, resulting in one electron leaving it completely (thus-forming a positively charged Chl molecule) and moving over to another molecule, which is then reduced, i.e., it becomes negatively charged. The detailed process differs between

36

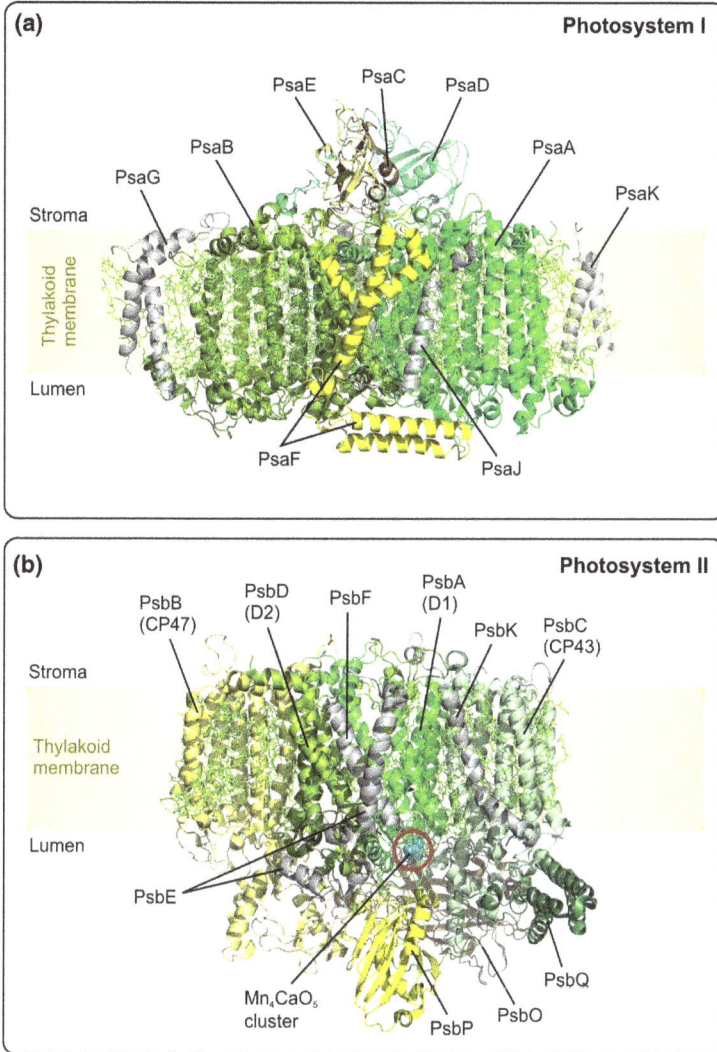

Fig. 3.5. Overall structures of higher plant Photosystem I (PSI) and Photosystem II (PSII) as viewed along the membrane. **(a)** Structure of PSI from pea (*Pisum sativum*) determined at 2.8 Å resolution. This PSI model was visualized using x-ray crystallographic coordinates deposited at Protein Data Bank (PDB) with ID 4Y28 [Mazor *et al.*, 2015]. **(b)** Structure of PSII monomer from spinach (*Spinacia oleracea*) obtained at 3.2 Å resolution. The PSII model was generated using cryo-EM data deposited at PDB with ID 3JCU [Wei *et al.*, 2016]. In both PSI and PSII structures major protein components have been colored. The location of the water-splitting Mn_4CaO_5 cluster (cyan balls) in PSII is highlighted by a red circle. See text and Figs. 3.6 and 3.7 for further details.

PSI and PSII, and is described in the legends of Figs 3.6, 3.7, and 3.8. Also see a review by Mamedov *et al.* [2015] for details of the primary photochemistry. We also note here that in addition to Chls *a*, other Chls may also participate in photochemical reactions. Nürnberg *et al.* [2018] have indicated that Chls *f* in reaction centers of some cyanobacteria are capable of absorbing and utilizing far-red light (around 750 nm) for driving charge separation.

The two photosystems are similar enough, and, thus, we believe that they must have evolved from a common ancestor, yet they exhibit large differences both in the overall protein structure and in the electron transfer chain of their reaction centers [Cardona, 2017]. It is highly likely that they diverged a very long time ago during evolution. According to an older theory, this divergent evolution took place in different organisms, which later fused into a single one, similar to recent cyanobacteria, which were incorporated into a nonphotosynthetic organism, already containing bacteria-derived mitochondria (see reviews by Olson and Blankenship [2004] and Blankenship [2010]). Even today, there are bacteria that carry out a primitive version of photosynthesis with a RC resembling PSI, while other bacteria have a RC resembling PSII, so one can imagine that a cyanobacterium containing two kinds of photosystems could have arisen by these two kinds of bacteria. But it is also possible [Allen, 2005; Allen and Martin, 2007] that they have diverged within a single kind of organism. Bacteria containing only one kind of photosystem would then have arisen by losing one of the photosystems. Gisriel *et al.* [2017] describe a homodimeric RC structure of a *Heliobacterium*, which provides a clue to the evolution, mentioned above.

One may wonder why plants and other organisms (cyanobacteria and algae) that carry out photosynthesis with evolution of oxygen by oxidation of water (Fig. 3.7) have two photosystems. As we shall see in the following section, the reason is the same as when we need two or more batteries (more correctly: two or more galvanic cells) to power, e.g., some torches, laser pointers, portable radios. It is because we need a sufficiently high potential difference between the two poles, and we get it by connecting our powering units in series. If you dip wires from the poles of a single 1.5 V battery into water that has been made conducting by the addition of, e.g., a little sulfuric acid, you do

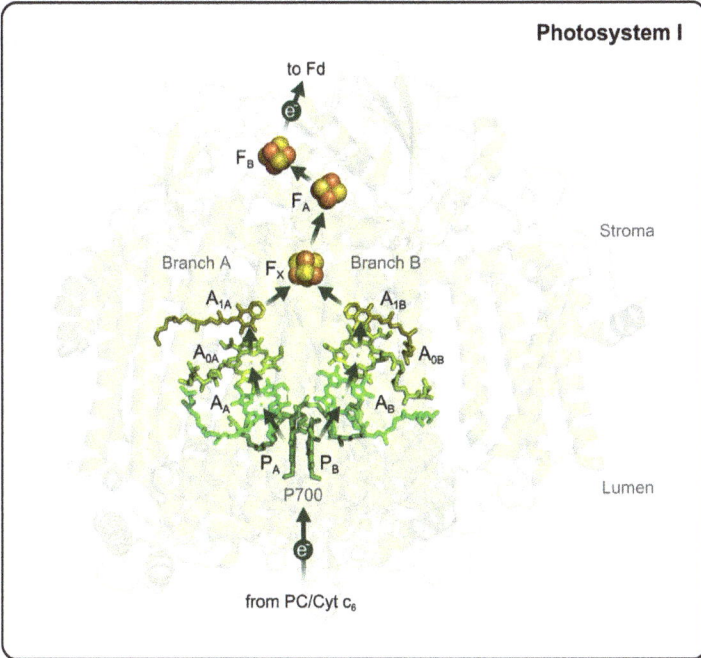

Fig. 3.6. Arrangement of electron-transfer cofactors in a reaction center of Photosystem I (PSI) as seen along the thylakoid membrane. The cofactors are shown in color and the protein skeleton as a light background. The electron transfer cofactors are organized in two parallel branches A (on the left) and B (on the right), named after the protein subunits PsaA and PsaB, respectively; see Fig. 3.5a. The arrows indicate the direction of the electron transfer and the two possible electron transfer pathways with equal probability (in contrast to the case in PSII). From antenna pigments (not shown) energy is transferred to a "special pair" of Chl *a* and Chl *a'* heterodimer (P_A and P_B), collectively referred to as P700 ("P" for pigment and "700" for its lowest energy absorption band at 700 nm). Excitation by transfer of a quantum (or an exciton: an "excited" electron) from the antenna pigments causes an electron to be transferred from P700* *via* the Chl molecules designated A and A_0 to the phylloquinone A_1. The six Chl molecules shown here are the only electron-transferring Chl molecules in PSI. In addition there are ~90 Chl *a* molecules (antenna) that serve only to absorb photons. While the first electron transfer, from Chl, occurs within 0.1 picoseconds (ps) [Shelaev *et al.*, 2010; also see Mamedov *et al.*, 2015], the electron transfer to A_1 takes ~25 ps and results in the formation of a radical pair, $P700^+A_1^-$. The electron is further transferred in sequence to the components labeled F_X, F_A, and F_B, which are iron-sulfur centers, and from F_B to ferredoxin (Fd; not shown), the soluble iron-sulfur protein in the stroma. (In some photosynthetic bacteria and algae, a "backup" flavin enzyme, flavodoxin, replaces Fd, under iron deficiency.) Oxidized P700 regains an electron from plastocyanin (PC), a small water-soluble copper protein, or from cytochrome c_6 (Cyt c_6) in cyanobacteria and some algae, in the lumen (not shown). The protein background and arrangement of the redox cofactors was modeled using coordinates deposited at PDB with ID 4Y28 [Mazor *et al.*, 2015].

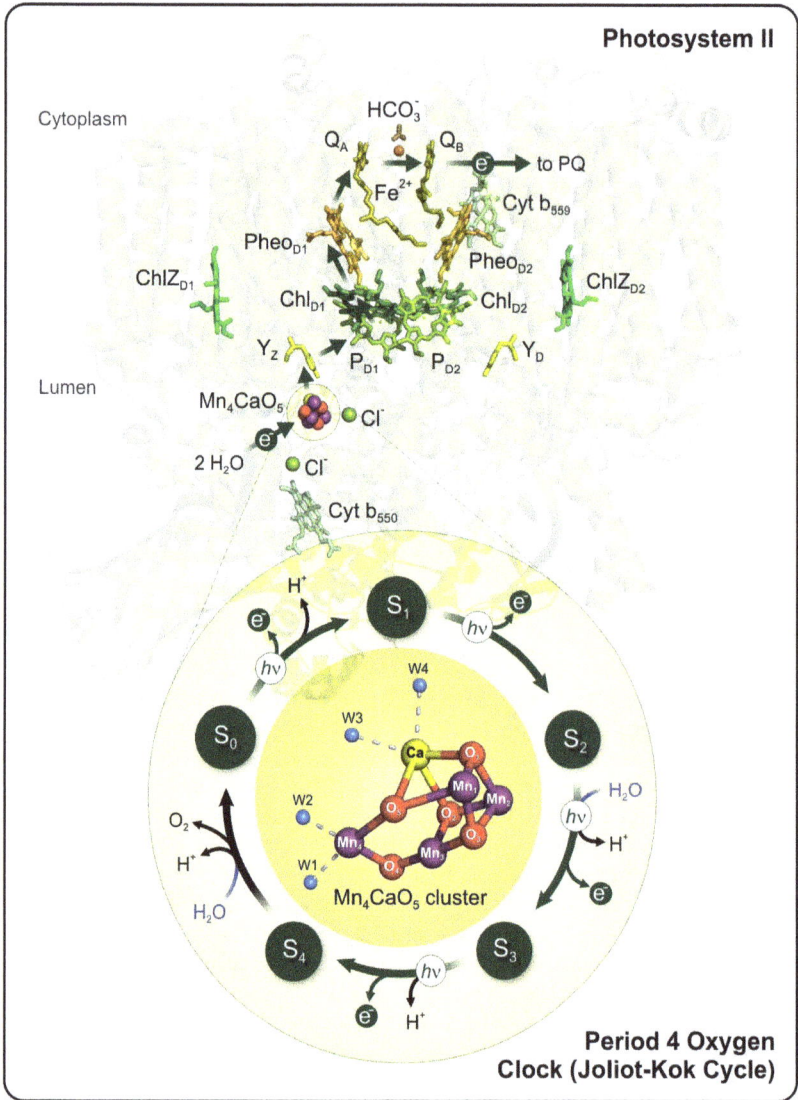

Photosystem II

Period 4 Oxygen Clock (Joliot-Kok Cycle)

(continued)

Fig. 3.7. (continued) Arrangement of the electron-transfer cofactors in the reaction center of Photosystem II (PSII; as seen along the thylakoid membrane; see the top part of the figure) and the Joliot-Kok* cycle (also known as the "oxygen clock"; see the bottom part of the figure) of photosynthetic water oxidation by the Mn_4CaO_5 cluster. The cofactors are shown in color and the protein scaffold as a light background. All redox cofactors of the PSII monomer are arranged in two branches on the D1 and D2 protein subunits (see Fig. 3.5b). The electron transfer (direction is indicated by arrows) occurs on the D1 protein of the PSII reaction center (*via* the so-called *active branch*). A quantum (or more appropriately, an "exciton" (an excited electron) is transferred from the antenna pigments, which ultimately leads to an electron to be transferred from the primary electron donor P680 (that refers to a pair of Chl *a* molecules (P_{D1} and P_{D2}) and two accessory Chls (Chl_{D1} and Chl_{D2})) to the pheophytin (Pheo), attached to the D1 polypeptide. From there it goes first to the plastoquinone Q_A on the D2 polypeptide, and then to the plastoquinone Q_B on the D1 polypeptide. This last transfer may involve the non-heme iron (Fe^{2+}; NHI), and the bicarbonate bound to it. Each of the plastoquinone molecules can take up, in sequential steps, two electrons, and the reduced Q_B takes up two protons; the fully reduced Q_BH_2, a plastoquinol, exchanges with a plastoquinone molecule (from the membrane) and is thus released from the Q_B-site of the D1 polypeptide into the lipid membrane [Van Eerden *et al.*, 2017]. Bicarbonate ion (HCO_3^-) plays an essential role here [Shevela *et al.*, 2012]. Y_Z and Y_D are tyrosine residues, of which Y_Z is used as an electron carrier between the Mn_4CaO_5 cluster of the oxygen-evolving complex (OEC) and P680. The structure of the Mn_4CaO_5 cluster, surrounded by four substrate water molecules (W1–W4) is shown within the Joliot-Kok cycle. The cycle consists of five S_i-states ($i = 0, ..., 4$), where *i* is the number of oxidizing equivalents stored within the Mn_4CaO_5 cluster of the OEC. Four of five redox-transitions between the S-states are driven by light (the $S_4 \rightarrow S_0$ transition does not require light). Thus, four light-induced removal of electrons to the Y_Z results in the release of four protons and one O_2 molecule from two water molecules per each turn of the S-state cycle. Note that depending on the S-state transition, either the proton or the electron is thought to be removed first. After a long dark adaptation of the photosynthetic system, the Mn_4CaO_5 cluster exists mostly in the S_1 state. For original version of the Joliot-Kok cycle, see Joliot *et al.* [1969] and Kok *et al.* [1970]. The arrangement of PSII redox cofactors, protein scaffold, and the structural model of the Mn_4CaO_5 cluster in the center of the cycle is based on the data of the x-ray crystallographic PSII structure obtained at 1.9 Å resolution [Umena *et al.*, 2011] deposited at PDB with ID 3ARC. We note that a detailed model of all the S-states will soon be available from the research group of Junko Yano; Vittal Yachandra; Jan Kern and their collaborates at Lawrence Berkeley Lab in California. [*Named after Bessel Kok and Pierre Joliot]

not get evolution of any gas, but if you connect two 1.5 V batteries in series you do. The positive pole has to be very positive to make oxygen in water give up two electrons per atom, i.e., four electrons per oxygen molecule formed.

3.3.2   Electron-transfer pathways

3.3.2.1   The "Z-scheme" of photosynthesis

The series connection of the two photosystems and the entire electron-transfer pathway of the light reactions of oxygenic photosynthesis are

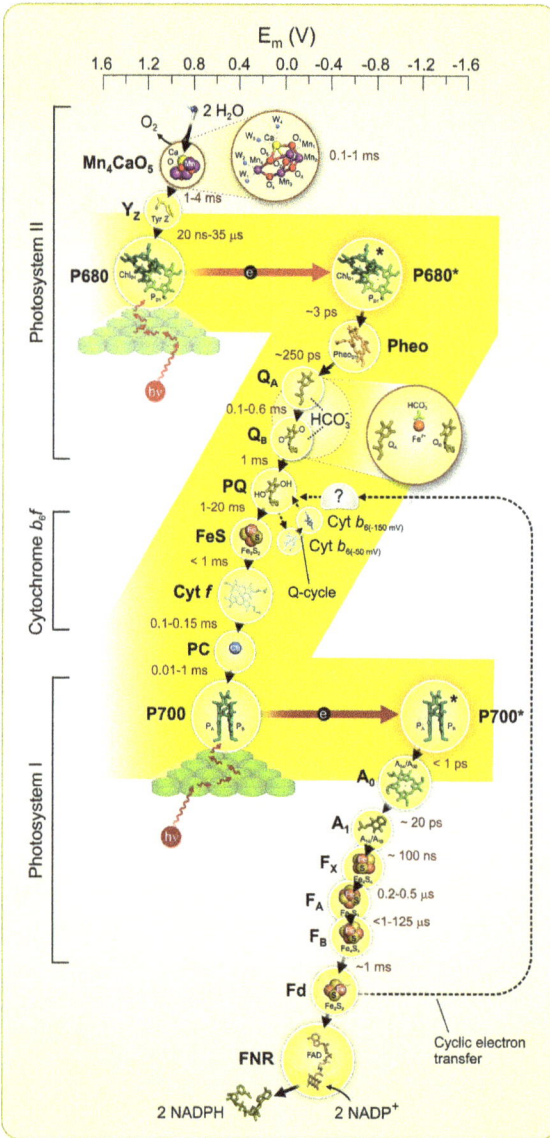

Fig. 3.8. The Z-Scheme for electron transport from water to NADP$^+$. The two photosystems, PSI and PSII are connected in series with redox components of the Cyt b_6f complex in between: FeS, the Rieske iron-sulfur protein; and Cyt f, cytochrome f. The transfer of electrons from PSII to the latter is mediated by plastoquinol (PQH$_2$) diffusing into the thylakoid lipid membrane, and the electron transfer from the Cyt b_6f complex to PSI by the highly mobile plastocyanin, PC, diffusing in the lumen. PQ represents a pool of mobile plastoquinone molecules. Several cyclic electron pathways, around PSI, have been suggested; for simplicity we present here only one, which may involve one or more proteins, shown with a question mark. The original version of this scheme, in its bare-bone form, was published in 1989 [Demeter and Govindjee, 1989], but the current scheme has been highly modified with the addition of structures of molecules in the scheme itself [cf. Govindjee *et al.*, 2017]. See legends of Figs. 3.6 and 3.7 for abbreviations; the approximate times for various steps are shown in brown text.

42

illustrated in Fig. 3.8. This type of diagram is called the *Z-scheme*, since the electron-transfer pathway plotted horizontally according to mid-point redox potentials of the cofactors at pH 7 (E_m 7) looks like the letter "Z." For historical overview on the Z-scheme, see Govindjee *et al.* [2013, 2017].

3.3.2.2   *Photosynthetic water oxidation (oxygen evolution)*

On the upper part of the Z-scheme (Fig. 3.8) we see water molecules, the source of the electrons for the electron-transfer chain of oxygenic photosynthesis. These water molecules are "split" (oxidized) into molecular oxygen (O_2), protons, and electrons (see Fig. 3.7), requiring specific end-products of the primary light reactions. The positive charge left on P680 after the "export" of an electron to pheophytin is neutralized by an electron coming from the tyrosine Y_Z, which in turn receives an electron from the Mn_4CaO_5 cluster of the oxygen-evolving complex (OEC) of PSII (see Figs. 3.7 and 3.8; also see [Vinyard *et al.,* 2013; Shen, 2015]). However, the formation of a molecule of oxygen requires the removal of four electrons from two water molecules. To understand how this is done, we must look more closely at Fig. 3.7. A clue as to how it functions was provided by experiments carried out first by Pierre Joliot *et al.* [1969] and then by Bessel Kok *et al.* [1970]. They found that a single short flash of light did not produce any molecular oxygen, no matter how strong the flash was. This is because several light dependent steps, each one requiring some time (longer than the duration of a single flash), are required. For the whole process to occur, the OEC undergoes a cyclic change, the S cycle, involving five separate states, from S_0 to S_4 (Fig. 3.7) (for reviews, see Cox and Messinger, 2013; Yano *et al.,* 2015; Pérez-Navarro *et al.,* 2016; Vinyard and Brudvig, 2017).

The OEC is located deep in the PSII protein complex, but the structure around it is spacious and provides a channel [Vogt *et al.,* 2015] (or even two [Zaraiskaya *et al.,* 2014]) through which the released oxygen molecule can escape rapidly into the lumen, and which also allows water to enter to replace what is used up. Since O_2 molecules are very soluble in lipids, they can easily penetrate the thylakoid membrane, the two membranes around the chloroplast, and the cell membrane, and then reach the intercellular air spaces in the leaves. From there they leave the organism (if it is a plant, through the stomata).

We still do not fully understand how the OEC is assembled during the life of a cyanobacterium or of a chloroplast (i.e., ontogenetically); further, how this center had evolved ~3 Ga ago is also a mystery. For current ideas on these issues, we refer the readers to Dismukes *et al.* [2001], Khorobrykh *et al.* [2013], Shevela *et al.* [2016], Khorobrykh *et al.* [2016], Bao and Burnap [2016], Cardona [2016], and Gates *et al.* [2017]; also see Eaton-Rye and Sobotka [2017] for further information.

3.3.2.3 Unique role of bicarbonate in light-induced reactions of PSII

Sixty years ago, Otto Heinrich Warburg, together with Günter Krippahl, discovered a unique stimulatory effect of CO_2 in the Hill reaction (i.e., O_2 evolution in the presence of artificial electron acceptors) [Warburg and Krippahl, 1958]. Warburg [1964] stated that this discovery supports his idea that O_2 in photosynthesis originates from CO_2. However, despite this mistaken interpretation, this discovery initiated a long-term debate about the possible role(s) and action site(s) of inorganic carbon, CO_2 and HCO_3^- (traditionally known as bicarbonate, or hydrogen carbonate, according to IUPAC) on photosynthetic O_2 evolution. Extensive research, over the last five decades, has clearly revealed that bicarbonate ions affect the electron flow on both the electron acceptor and the donor side of PSII (see reviews by Blubaugh and Govindjee [1988], Van Rensen *et al.* [1999], Stemler [2002], Van Rensen and Klimov [2005], McConnell *et al.* [2012], Shevela *et al.* [2012]). On the electron acceptor side, bicarbonate is bound to non-heme iron (NHI) between Q_A and Q_B, the two plastoquinone electron acceptors (Figs. 3.7 and 3.8), and that this bicarbonate is essential for electron and proton transfer at the Q_B-site, as originally discovered by Wydrzynski and Govindjee [1975], and shown convincingly by Govindjee *et al.* [1976] that the effect is on the "two-electron gate," i.e., on the Q_B-site [Velthuys and Amesz, 1974].

Since 1970s bicarbonate has also been suggested to act on the water-splitting (donor) side of PSII [Stemler *et al.*, 1974; Stemler, 2002].

Later, it was shown that bicarbonate ions have several effects on this side: as a mobile proton acceptor [Ananyev *et al.*, 2005; Shutova *et al.*, 2008; Koroidov *et al.*, 2014; Benlloch *et al.*, 2015], as a native cofactor in the photoassembly of the Mn_4CaO_5 cluster [Baranov *et al.*, 2004; Dasgupta *et al.*, 2008], or as a stabilizing agent for the activity of the OEC [Klimov *et al.*, 1997; Van Rensen and Klimov, 2005]. Despite these multiple bicarbonate effects within PSII, recent x-ray crystallographic and cryo-EM studies have clearly shown that only one HCO_3^- ion firmly binds to PSII (as a bidentate ligand to NHI between Q_A and Q_B; see Figs. 3.7 and 3.8) of cyanobacteria, algae, and higher plants [Umena *et al.*, 2011; Ago *et al.*, 2016; Wei *et al.*, 2016]. Interestingly, while this bound bicarbonate has been observed in RCII of all oxygenic organisms, it is not present in the RCs of anoxygenic photosynthetic bacteria, as it was noticed already more than 30 years ago [Michel and Deisenhofer, 1988; Shopes *et al.*, 1989; Wang *et al.*, 1992]. It is obvious, therefore, that by being a ligand to the NHI between Q_A and Q_B, and binding to amino acids of the D1 and D2 proteins of PSII (Fig. 3.7; also see Fig. 16 in a review by Shevela *et al.* [2012]), bicarbonate plays a unique role only in oxygenic photosynthesis: it stabilizes the Q_A–NHI–Q_B structure of the PSII RC, allowing protonation of Q_B^- *via* certain amino acids around this site [Xiong *et al.*, 1998; Müh *et al.*, 2012; Shevela *et al.*, 2012] and, thereby, accelerates electron transfer from the reduced Q_A to Q_B and from Q_B into PQ-pool. Further research is needed to fully decipher the mechanisms involved on the functioning of bicarbonate on both sides of PSII.

3.3.2.4   Formation of reducing power

A part of the Z-scheme (Fig. 3.8) that remains to be described further is the bottom part, in which the electrons leave the membrane-bound PSI to be taken up by the water-soluble iron-sulfur protein ferredoxin (Fd) in the stroma. The reduced Fd is reoxidized by the flavin enzyme ferredoxin-nicotinamide nucleotide phosphate oxidoreductase (FNR), which transfers the electrons to $NADP^+$. The doubly reduced nicotinamide nucleotide phosphate (NADPH) provides reducing power to the carbon dioxide assimilation, which will be described in Chapter 4.

3.3.3 *Proton-transfer pathways and formation of ATP (photophosphorylation)*

Thus far, we have dealt with the transfer of electrons from water to the reducing end of the Z-scheme. But this electron "flow" is intertwined with the consumption and the release of protons, which results in a net cyclic flow of protons.

To start with the oxidation of water: Removal of electrons and release of oxygen leaves protons behind on the lumen side of the thylakoid membrane. When two electrons are transferred to the plastoquinone Q_B at the reducing end of PSII, two protons are taken up from the stroma, ultimately to produce plastoquinol (plastohydroquinone; Fig. 3.9) [Van Eerden *et al.*, 2017].

The plastoquinol is indeed reoxidized by the Cyt b_6f complex in a remarkable way, giving off two protons to the lumen. One of the electrons available from this process is used to reduce Cyt f, while the other one, *via* the Cyt b, contributes to the re-reduction of one plastoquinone molecule attached to a binding site on the Cyt b_6f complex (Fig. 3.10). Thus, for every electron continuing to PSI, two protons are transported from the stroma to the lumen. Together with what happened in PSII, the transport of two electrons results in the transport of six electrons from the stroma to the lumen. Finally, at the reducing (electron acceptor) end of PSI, two electrons reducing $NADP^+$ to NADPH need the uptake of one proton, but this is set free again as NADPH is oxidized during the process

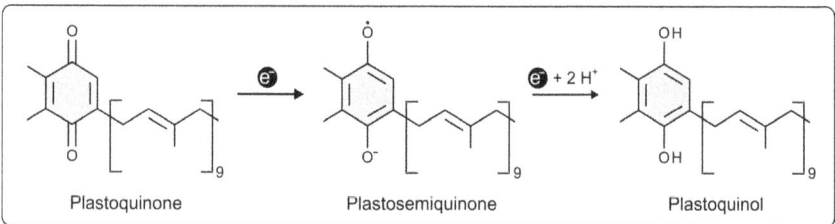

Fig. 3.9. Structures and reactions of plastoquinone, plastosemiquinone, and plastoquinol on the electron-acceptor side of PSII. Reduction of plastoquinone by one electron results in the formation of plastosemiquinone, and then, after the addition of one more electron and two protons, plastoquinol is formed.

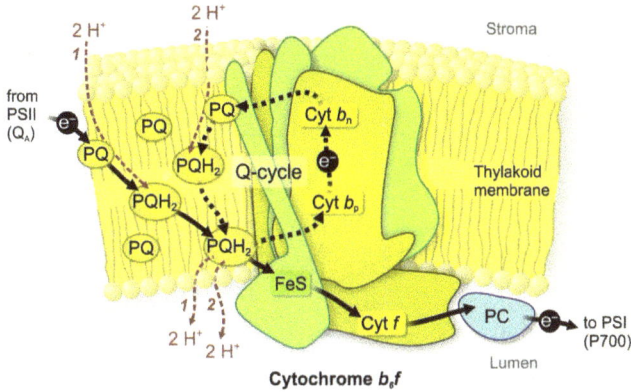

Fig. 3.10. Schematic representation of the linear electron-transfer (solid arrows) and the "Q-cycle" (dashed arrows) *via* the Cyt b_6f complex. Proton translocation involved in this electron-transfer reaction is shown by brown dashed arrows. When PQH_2 from PSII arrives at the Cyt b_6f complex, it reduces the FeS protein, and protons are released into the lumen (arrows marked as "1"). One of two available electrons delivered by the PQH_2 goes to the Cyt f subunit and further to PC, which then carries the electrons towards PSI. The other electron, after being transferred to hemes of two kinds of Cyt b (labeled as b_p and b_n for the one on the positive and negative side of the membrane; cf. Fig. 3.8 where they are labeled with their midpoint redox potentials); the electron on Cyt b_n then reduces a PQ (arrows marked as "2"). When PQ is reduced twice it picks up two protons (and forms PQH_2), participating thereby in the Q-cycle. This Q-cycle increases the number of protons pumped across the thylakoid membrane [Mulkidjanian, 2010; Cramer *et al.*, 2011].

of CO_2-assimilation; there is no net change in protonation. Thus the proton/electron ratio is 6/2 = 3. As we shall see, this ratio can be modified by the "cyclic electron flow" around PSI (see Fig. 3.11). Cyclic electron flow results in a transport of protons without any net transport of electrons and production of reducing power. The role of cyclic electron transport in higher plants and its importance in enabling flexibility, adjustment to environmental conditions, and protection, has recently attracted considerable interest [Johnson, 2011; Shikanai, 2014; Huang *et al.*, 2015; Suorsa, 2015; Wang *et al.*, 2015; Yamori *et al.*, 2015].

The transfer of protons from the stroma to the lumen cannot continue without being balanced in some way, and it is here that the fourth large protein in the thylakoid membrane, the *ATP synthase*, comes in (Fig. 3.12a). The disequilibrium of protons in stroma and lumen developed

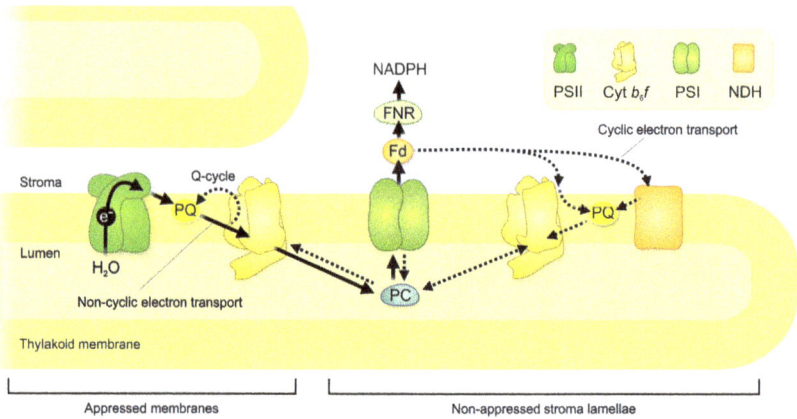

Fig. 3.11. A simplified view of the Z-scheme (non-cyclic electron flow) shown together with cyclic electron transport. The left side, "Appressed membranes," shows the part of a thylakoid membrane, which is in a granum, whereas the right side ("Non-appressed stroma lamellae") shows a membrane of a thylakoid surrounded by stroma. In non-cyclic electron transport (solid arrows), electrons flow from water *via* PSII, PQ, the Cyt b_6f complex, PC and PSI to Fd. In cyclic electron transport (dashed arrows), electrons flow from PSI *via* Fd and either of two proteins (PGR5 (not shown) or NDH), PQ, the Cyt b_6f complex and PC back to PSI. The cyclic electron transport provides additional translocation of protons from the stroma to the lumen without any net electron transport. Modified from Johnson [2011].

by the proton transport described above represents a store of energy that is used by the ATP synthase to make ATP (adenosine triphosphate), using ADP (adenosine diphosphate), and inorganic orthophosphate (Pi). Such synthesis is not unique for photosynthetic processes, but occurs also in bacteria and mitochondria during respiration, and the ATP synthase is of very ancient evolutionary origin [Mulkidjanian *et al.*, 2007], yet has a very complex structure. ATP synthase has been studied by numerous researchers, among those Paul Boyer, who explained how this enzyme functions (by his famous "binding change (rotation) mechanism"), and John Walker, who obtained its X-ray crystallography structure; the two received the Nobel Prize in Chemistry in 1997 "for their elucidation of the enzymatic mechanism underlying the synthesis of ATP" [Abrahams *et al.*, 1994; Boyer, 1997; also see Walker's web site http://www.mrc-mbu.cam.ac.uk/people/john-walker]. We know, that this enzyme consists of a "stator" fixed in the thylakoid membrane, and a "rotor" consisting of two parts connected by

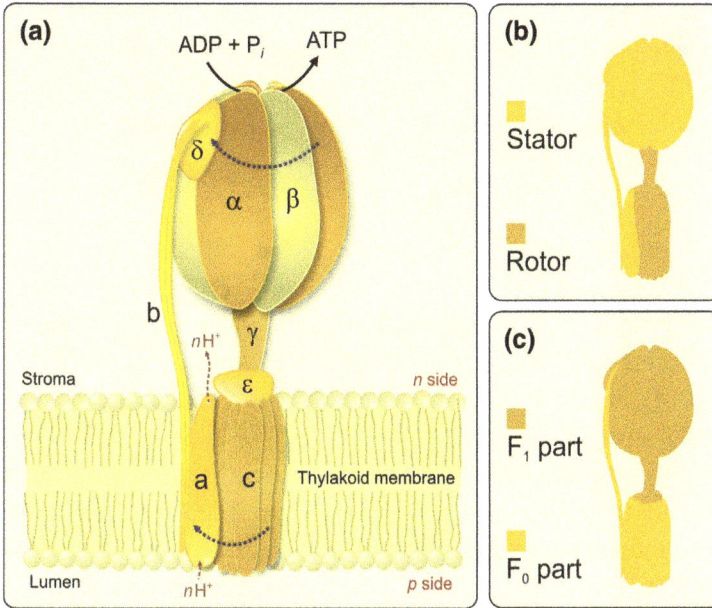

Fig. 3.12. Schematic representation of ATP synthase and its subunits. **(a)** Overview of the ATP synthase subunits. **(b)** Schematic view of "rotor" and "stator" subunits of ATP synthase (shaded differently as shown in the figure). **(c)** Schematic representation of F_o and F_1 parts of ATP synthase (shaded differently as shown in the figure). Modified from Shevela *et al.* [2013]

an elastic membrane (Fig. 3.12b). Protons move through the ATP synthase *via* a channel from the lumen to the stroma, driven by the difference in proton concentration between the two compartments, as well as an electrical potential difference associated with it; a part of the rotor, the F_o, rotates in a stepwise manner, dragging the other part of the rotor (F_1) around (Fig. 3.12c). The rotation of the two parts of the rotor are not completely in step, and the elasticity of the connecting shaft (the main part of which is the γ subunit) makes this "out of step" possible (Fig. 3.12a). The importance of fine-tuned mechanical properties of the shaft for energy conservation is discussed by Okazaki and Hummer [2015]; the review by Mukherjee and Warshel [2017] provides further details and new concepts on the structure and function of the F_oF_1 ATP synthase.

The substrates, ADP and orthophosphate (P_i), become bound to sites on F_1, and when F_1 is "forced" to rotate, the regions around these binding sites undergo a conformational change, which forces the substrates together and "squeezes" out the water:

$$ADP + P_i \rightarrow ATP + H_2O \qquad (3.2)$$

The subunit epsilon (ε) is thought to be important for preventing rehydrolysis of ATP [Bockenhauer *et al.*, 2014]. The operation of the ATP synthase is regulated in several ways, mainly *via* subunit γ [Sunamura *et al.*, 2012; Buchert *et al.*, 2015] and subunit ε [Mizumoto *et al.*, 2013; Duncan *et al.*, 2014]. Redox regulation is achieved *via* the so-called thioredoxin system [Hisabori *et al.*, 2013].

The "*chemiosmotic hypothesis*," where phosphorylation is powered by a difference of ionic concentration and of electrical potential across biological membranes, was proposed by Peter Mitchell in 1961 and substantiated by 1966; Mitchell was awarded the Nobel Prize for Chemistry in 1978 for this concept. (See a republished version [Mitchell, 2011].). A direct demonstration that phosphorylation in chloroplasts is driven by a difference in proton concentration across the thylakoid membrane was provided by Jagendorf and Uribe [1966] (also see [Jagendorf, 1967]). In this experiment, chloroplast fragments were first kept at low pH, and then a proton concentration difference across the thylakoid membranes was given by a sudden transfer to high pH; this led to ATP synthesis without the use of light. Soon after another experiment demonstrated that whole thylakoid membrane sacs containing ~100,000 Chl molecules act as a unit for energy storage [Junge *et al.*, 1968; Junge and Witt, 1968; Björn, 1971a,b]. Furthermore, the concept that membrane potential, the other component of "proton motive force," the first being proton gradient, can indeed make ATP was discussed, e.g., by Damroth *et al.* [2000]. All of the above are in agreement with the modern view of ATP synthesis. However, we shall keep our minds open for alternative views.

In Chapter 4 we shall talk about CO_2 fixation pathways in photosynthesis.

References

Abrahams, J. P., Leslie, A. G., Lutter, R. and Walker J. E. (1994). Structure at 2.8 Å resolution of F 1-ATPase from bovine heart mitochondria, *Nature*, 370, pp. 621–628.

Ago, H., Adachi, H., Umena, Y., Tashiro, T., Kawakami, K., Kamiya, N., Tian, L., Han, G., Kuang, T., Liu, Z., Wang, F., Zou, H., Enami, I., Miyano, M. and Shen, J.-R. (2016). Novel features of eukaryotic photosystem II revealed by its crystal structure analysis from a red alga, *J. Biol. Chem.*, 291, pp. 5676–5687.

Allen, J. F. (2005). A redox switch hypothesis for the origin of two light reactions in photosynthesis, *FEBS Lett.*, 579, pp. 963–968.

Allen, J. F. and Martin, W. (2007). Evolutionary biology – Out of thin air, *Nature*, 445, pp. 610–612.

Ananyev, G., Nguyen, T., Putnam-Evans, C. and Dismukes, G. C. (2005). Mutagenesis of CP43-arginine-357 to serine reveals new evidence for (bi)carbonate functioning in the water oxidizing complex of photosystem II, *Photochem. Photobiol. Sci.*, 4, pp. 991–998.

Atkins, P. W. and Friedman, R. S. (1999). *Molecular Quantum Mechanics* (Oxford University Press, Oxford).

Bao, H. and Burnap, R. L. (2016). Photoactivation: The light-driven assembly of the water oxidation complex of Photosystem II, *Front. Plant Sci.*, 7, p. 578.

Baranov, S. V., Tyryshkin, A. M., Katz, D., Dismukes, G. C., Ananyev, G. M. and Klimov, V. V. (2004). Bicarbonate is a native cofactor for assembly of the manganese cluster of the photosynthetic water oxidizing complex. Kinetics of reconstitution of O_2 evolution by photoactivation, *Biochemistry*, 43, pp. 2070–2079.

Bendall, D. S. and Manasse, R. S. (1995). Cyclic photophosphorylation and electron transport, *Biochim. Biophys. Acta*, 1229, pp. 23–38.

Benlloch, R., Shevela, D., Hainzl, T., Grundström, C., Shutova, T., Messinger, J., Samuelsson, G. and Sauer-Eriksson, A. E. (2015). Crystal structure and functional characterization of photosystem II-associated carbonic anhydrase CAH3 in *Chlamydomonas reinhardtii*, *Plant Physiol.*, 167, pp. 950–962.

Björn, L. O. (1971a). Effects of some chemical compounds on the slow, far-red induced afterglow from leaves of *Elodea* and the lack of effect of N-methyl phenazonium methosulfate on far-red induced glucose uptake in *Chlorella*, *Physiol. Plant.*, 25, pp. 316–323.

Björn, L. O. (1971b). Far-red induced, long-lived afterglow from photosynthetic cells. Size of afterglow unit and paths of energy accumulation and dissipation, *Photochem. Photobiol.*, 13, pp. 5–20.

Blankenship, R. E. (2010). Early evolution of photosynthesis, *Plant Physiol.*, 154, pp. 434–438.

Blubaugh, D. J. and Govindjee. (1988). The molecular mechanism of the bicarbonate effect at the plastoquinone reductase site of photosynthesis, *Photosynth. Res.*, 19, pp. 85–128.

Bockenhauer, S. D., Duncan, T. M., Moernera, W. E. and Börsch, M. (2014). The regulatory switch of F1-ATPase studied by single-molecule FRET in the ABEL Trap, *Proc. SPIE*, 8950, p. 89500H.

Boyer, P. D. (1997). The ATP synthase — a splendid molecular machine, *Annu. Rev. Biochem.*, 66, pp. 717–749.

Buchert, F., Konno, H. and Hisabori, T. (2015). Redox regulation of CF1-ATPase involves interplay between the γ-subunit neck region and the turn region of the βDELSEED-loop, *Biochim. Biophys. Acta*, 1847, pp. 441–450.

Cardona, T. (2016). Reconstructing the origin of oxygenic photosynthesis: Do assembly and photoactivation recapitulate evolution? *Front. Plant Sci.*, 7, p. 257.

Cardona, T. (2017). Evolution of photosynthesis. *In eLS* (John Wiley & Sons Ltd, Chichester), doi: 10.1002/9780470015902.a0002034.pub3

Cramer, W. A., Hasan, S. S. and E. Yamashita (2011). The Q cycle of cytochrome *bc* complexes: A structure perspective, *Biochim. Biophys. Acta*, 1807, pp. 788–802.

Cox, N. and Messinger, J. (2013). Reflections on substrate water and dioxygen formation, *Biochim. Biophys. Acta*, 1827, pp. 1020–1030.

Damroth, P., Kaim, G. and Matthey, U. (2000). Crucial role of the membrane potential for ATP synthesis by F1 Fo ATP synthases, *J. Exp. Biol.*, 203, pp. 51–59.

Dasgupta, J., Ananyev, G. M. and Dismukes, G. C. (2008). Photoassembly of the water-oxidizing complex in photosystem II, *Coord. Chem. Rev.*, 252, pp. 347–360.

Dean, J. C., Mirkovic, T., Toa, S. D., Oblinsky, D. G. and Scholes, D. G. (2016). Vibronic enhancement of algae light harvesting, *Chem*, 1, pp. 858–872.

Demeter, S. and Govindjee (1989). Thermoluminescence in plants, *Physiol. Plant.*, 75, pp. 121–130.

Dismukes, G. C., Klimov, V. V., Baranov, S. V., Kozlov, Y. N., DasGupta, J. and Tyryshkin, A. (2001) The origin of atmospheric oxygen on Earth: The innovation of oxygenic photosynthesis, *Proc. Natl Acad. Sci. U.S.A.*, 98, pp. 2170–2175.

Duncan, T. M., Düser, M. G., Heitkamp, T., McMillan, D. G. G. and Börsch, M. (2014). Regulatory conformational changes of the ε subunit in single FRET-labeled F(o)F(1)-ATP synthase, *Proc. SPIE Int. Soc. Opt. Eng.*, 8948, p. 89481J.

Eaton-Rye, J. J. and Sobotka, R., eds. (2017). *Assembly of the Photosystem II Membrane-Protein Complex of Oxygenic Photosynthesis* (Frontiers Media, Lausanne). doi:10.3389/978-2-88945-233-0

Gates, C., Ananyev, G. and Dismukes, G. C. (2017). Photoassembly of the $CaMn_4O_5$ catalytic core and its inorganic mutants in photosystem II. *In Encyclopedia of Inorganic and Bioinorganic Chemistry* (John Wiley & Sons, Ltd.), p. 15. doi:10.1002/9781119951438.eibc2481

Gisriel, C., Sarrou, I., Ferlez, B., Golbeck, J. H., Redding, K. E. and Fromme, R. (2017). Structure of a symmetric photosynthetic reaction center-photosystem, *Science*, 357, pp. 1021–1025.

Govindjee, Björn, L. O. and Nickelsen, K. (2013). Evolution of the Z-scheme of electron transport in oxygenic photosynthesis. *In Photosynthesis Research for Food, Fuel and the Future: 15th International Conference on Photosynthesis* (Springer, Heidelberg), pp. 827–833.

Govindjee, Pulles, M. P. J., Govindjee, R., Van Gorkom, H. J. and Duysens, L. N. M. (1976). Inhibition of the reoxidation of the secondary electron acceptor of photosystem II by bicarbonate depletion, *Biochim. Biophys. Acta*, 449, pp. 602–605.

Govindjee, Shevela, D. and Björn, L. O. (2017). Evolution of the Z-scheme of photosynthesis: A perspective, *Photosynth. Res.*, 133, pp. 5–15.

Hisabori, T., Sunamura, E.-I., Kim, Y. and Konno, H. (2013). The chloroplast ATP synthase features the characteristic redox regulation machinery, *Antioxid. Redox Signal.*, 19, pp. 1846–1854.

Huang, W., Yang, Y.-J., Hu, H. and Zhang, S.-B. (2015). Different roles of cyclic electron flow around photosystem I under sub-saturating and saturating light intensities in tobacco leaves, *Front. Plant Sci.*, 6, p. 923.

Jagendorf, A. T. (1967). Acid-base transitions and phosphorylation by chloroplasts, *Fed. Proc.*, 26, pp. 1361–1369.

Jagendorf, A. T. and Uribe, E. (1966). ATP formation caused by acid-base transitions of spinach chloroplasts. *Proc. Natl Acad. Sci. U.S.A.*, 55, pp. 170–177.

Johnson, G. N. (2011). Physiology of PSI cyclic electron transport in higher plants, *Biochim. Biophys. Acta*, 1807, pp. 384–389.

Joliot, P., Barbieri, G. and Chabaud, R. (1969). Un nouveau modele des centres photochimiques du systeme II, *Photochem. Photobiol.*, 10, pp. 309–329.

Joliot, P., Joliot, A. and Johnson, G. (2006). Cyclic electron transfer around photosystem I. *In Golbeck, J. H., ed, Photosystem I: The Light-Driven Plastocyanine: Ferredoxin Oxidoreductase.* (Springer, Dordrecht), pp. 639–656.

53

Junge, W., Reinwald, E., Rumberg, B., Siggel, U. and Witt, H. T. (1968). Further evidence for a new function unit of photosynthesis, *Naturwissenschaften*, 55, pp. 36–37.

Junge, W. and Witt, H. T. (1968). On the ion transport system of photosynthesis – Investigations on a molecular level, *Z. Naturforsch. B*, 23, pp. 244–254.

Klimov, V. V., Baranov, S. V. and Allakhverdiev, S. I. (1997). Bicarbonate protects the donor side of photosystem II against photoinhibition and thermoinactivation, *FEBS Lett.*, 418, pp. 243–246.

Kok, B., Forbush, B. and McGloin, M. (1970). Cooperation of charges in photosynthetic O_2 evolution, *Photochem. Photobiol.*, 11, pp. 457–476.

Khorobrykh, A., Dasgupta, J., Kolling D. R., Terentyev, V., Klimov V., V. and Dismukes, G. C. (2013). Evolutionary origins of the photosynthetic water oxidation cluster: Bicarbonate permits Mn^{2+} photo-oxidation by anoxygenic bacterial reaction centers, *ChemBioChem*, 14, pp. 1725–1731.

Khorobrykh, A. A., Yanykin, D. V. and Klimov V. V. (2016). Enhancement of photoassembly of the functionally active water-oxidizing complex in Mn-depleted photosystem II membranes upon transition to anaerobic conditions, *J. Photochem. Photobiol. B*, 163, pp. 211–215.

Koroidov, S., Shevela, D., Shutova, T., Samuelsson, G. and Messinger, J. (2014). Mobile hydrogen carbonate acts as proton acceptor in photosynthetic water oxidation, *Proc. Natl. Acad. Sci. U.S.A.*, 111, pp. 6299–6304.

Kurisu, G., Zhang, H., Smith, J. L. and Cramer, W. A. (2003). Structure of the cytochrome b_6f complex of oxygenic photosynthesis: Tuning the cavity, *Science*, 302, pp. 1009–1014.

Mamedov, M., Govindjee, Nadtochenko, V. and Semenov, A. (2015). Primary electron transfer processes in photosynthetic reaction centers from oxygenic organisms, *Photosynth. Res.*, 125, pp. 51–63.

Mazor, Y., Borovikova, A. and Nelson, N. (2015). The structure of plant photosystem I super-complex at 2.8 Å resolution, *eLife*, 4, p. e07433.

McConnell, I. L. Eaton-Rye, J. J. and Van Rensen, J. J. S. (2012). Regulation of photosystem II electron transport by bicarbonate. In: Eaton-Rye, J. J., Tripathy, B. C. and Sharkey, T. D., eds, *Photosynthesis: Plastid Biology, Energy Conversion and Carbon Assimilation* (Springer, Dordrecht), pp. 475–500.

Mirkovic, T., Ostrumov, E. E., Anna, J. M., Van Grondelle, R., Govindjee and Scholes, G. D. (2017). Light absorption and energy transfer in the antenna complexes of photosynthetic organisms. *Chem. Rev.*, 117, pp. 249–293.

Michel, H. and Deisenhofer J. (1988). Relevance of the photosynthetic reaction center from purple bacteria to the structure of photosystem II, *Biochemistry*, 27, pp. 1–7.

Mitchell, P. (2011). Chemiosmotic coupling in oxidative and photosynthetic phosphorylation, *Biochim. Biophys. Acta*, 1807, pp. 1507–1538.

Mizumoto, J., Kikuchi, Y., Nakanishi, Y.-H., Mouri, N., Cai, A., Ohta, T., Haruyama, T. and Kato-Yamada, Y. (2013). ε Subunit of *Bacillus subtilis* F1-ATPase relieves MgADP inhibition, *PLoS ONE*, 8, p. e73888.

Mukherjee, S. and Warshel, A. (2017). The F_OF_1 ATP synthase: From atomistic three-dimensional structure to the rotary-chemical function, *Photosynth. Res.*, 134, pp. 1–15.

Mulkidjanian, A. Y., Makarova, K. S., Galperin, M. Y. and Koonin, E. V. (2007). Inventing the dynamo machine: The evolution of the F-type and V-type ATPases, *Nat. Rev. Microbiol.*, 5, pp. 892–899.

Mulkidjanian, A. Y. (2010) Activated Q-cycle as a common mechanism for cytochrome bc_1 and cytochrome b_6f complexes, *Biochim. Biophys. Acta*, 1797, pp. 1858–1868.

Müh, F., Glöckner, C., Hellmich, J. and Zouni, A. (2012). Light-induced quinone reduction in photosystem II, *Biochim. Biophys. Acta*, 1817, pp. 44–65.

Nürnberg, D. J., Morton, J., Santabarbara, S., Telfer, A., Joliot, P., Antonaru, L. A., Ruban, A. V., Cardona, T., Krausz, E., Boussac, A., Fantuzzi, A. and Rutherford, A. W. (2018). Photochemistry beyond the red limit in chlorophyll f–containing photosystems, *Science*, 360, pp. 1210–1213.

Okazaki, K.-I. and Hummer, G. (2015). Elasticity, friction, and pathway of γ-subunit rotation in FoF1-ATP synthase, *Proc. Natl. Acad. Sci. U.S.A.*, 112, pp. 10720–10725.

Olson, J. M. and Blankenship, R. E. (2004). Thinking about the evolution of photosynthesis, *Photosynth. Res.*, 80, pp. 373–386.

Pérez-Navarro, M., Neese, F., Lubitz, W., Pantazis, D. A. and Cox, N. (2016). Recent developments in biological water oxidation, *Curr. Opin. Chem. Biol.*, 31, pp. 113–119.

Rabinowitch, E. and Govindjee. (1969). *Photosynthesis* (John Wiley & Sons Inc., New York).

Reimers, J. R., Cai, Z.-L., Kobayashi, R., Rätsep, M., Freiberg, A. and Krausz, E. (2013). Chlorophylls: Coherence loss *via* Q_x–Q_y mixing, *Sci. Rep.*, 3, p. 2761.

Shelaev, I. V., Gostev, F. E., Mamedov, M. D., Sarkisov, O. M., Nadtochenko, V. A., Shuvalov, V. A. and Semenov, A. Y. (2010). Femtosecond primary charge separation in Synechocystis sp. PCC 6803 photosystem I, *Biochim. Biophys. Acta*, 1797, pp. 1410–1420.

Shen, J.-R. (2015). The structure of photosystem II and the mechanism of water oxidation in photosynthesis, *Annu. Rev. Plant Biol.*, 66, pp. 23–48.

Shevela, D., Eaton-Rye, J. J., Shen, J.-R. and Govindjee. (2012). Photosystem II and the unique role of bicarbonate: A historical perspective, *Biochim. Biophys. Acta*, 1817, pp. 1134–1151.

Shevela, D., Pishchalnikov, R. Y., Eichacker, L. A. and Govindjee. (2013). Oxygenic photosynthesis in cyanobacteria. *In* Srivastava, A. K., Amar, N. R., Neilan, B. A., eds, *Stress Biology of Cyanobacteria: Molecular Mechanisms to Cellular Responses* (CRC Press/Taylor & Francis Group, Boca Raton, FL), pp. 3–40.

Shevela, D., Arnold, J., Reisinger, V., Berends, H.-M., Kmiec, K., Koroidov, S., Bue, A. K., Messinger, J. and Eichacker, L. A. (2016). Biogenesis of water splitting by photosystem II during de-etiolation of barley (*Hordeum vulgare* L.), *Plant Cell Environ.*, 39, pp. 1524–1536.

Shikanai, T. (2014). Central role of cyclic electron transport around photosystem I in the regulation of photosynthesis, *Curr. Opin. Biotechnol.*, 26, pp. 25–30.

Shopes, R. J., Blubaugh, D. J., Wraight, C. A. and Govindjee. (1989). Absence of a bicarbonate-depletion effect in electron transfer between quinones in chromatophores and reaction centers of *Rhodobacter sphaeroides*, *Biochim. Biophys. Acta*, 974, pp. 114–118.

Shutova, T., Kenneweg, H., Buchta, J., Nikitina, J., Terentyev, V., Chernyshov, S., Andersson, B., Allakhverdiev, S. I., Klimov, V. V., Dau, H., Junge, W. and Samuelsson, G. (2008). The photosystem II-associated Cah3 in *Chlamydomonas* enhances the O_2 evolution rate by proton removal, *EMBO J.*, 27, pp. 782–791.

Stemler, A., Babcock, G. T. and Govindjee. (1974). Effect of bicarbonate on photosynthetic oxygen evolution in flashing light in chloroplast fragments, *Proc. Natl. Acad. Sci. U.S.A.*, 71, pp. 4679–4683.

Stemler, A. J. (2002). The bicarbonate effect, oxygen evolution, and the shadow of Otto Warburg, *Photosynth. Res.*, 73, pp. 177–183.

Sunamura, E.-I., Konno, H., Imashimizu, M., Mochimaru, M. and Hisabori, T. (2012). A conformational change of the γ subunit indirectly regulates the activity of cyanobacterial F(1)-ATPase, *J. Biol. Chem.*, 287, pp. 38695–38704.

Suorsa, M. (2015). Cyclic electron flow provides acclimatory plasticity for the photosynthetic machinery under various environmental conditions and developmental stages. *Front. Plant Sci.*, 6, 800.

Umena, Y., Kawakami, K., Shen, J.-R. and Kamiya, N. (2011). Crystal structure of oxygen-evolving photosystem II at a resolution of 1.9 Å, *Nature*, 473, pp. 55–60.

Van Eerden, F. J., Melo, M. N., Frederix, P. W. J. M., Periole, X. and Marrink, S. J. (2017). Exchange pathways of plastoquinone and plastoquinol in the photosystem II complex, *Nat. Comm.*, 8, p. 15214.

Velthuys, B. R. and Amesz, J. (1974). Charge accumulation at the reducing side of system 2 of photosynthesis, *Biochim. Biophys. Acta*, 333, pp. 85–94.

Vinyard, D. J., Ananyev, G. M. and Dismukes, G. C. (2013). Photosystem II: The reaction center of oxygenic photosynthesis, *Annu. Rev. Biochem.*, 82, pp. 577–606.

Vinyard, D. J. and Brudvig, G. W. (2017). Progress toward a molecular mechanism of water oxidation in photosystem II, *Annu. Rev. Phys. Chem.*, 68, pp. 101–116.

Vogt, L., Vinyard, D. J., Khan, S. and Brudvig, G. W. (2015). Oxygen-evolving complex of Photosystem II: An analysis of second-shell residues and hydrogen-bonding networks, *Curr. Opin. Chem. Biol.*, 25, pp. 152–158.

Wang, C., Yamamoto, H. and Shikanai, T. (2015). Role of cyclic electron transport around photosystem I in regulating proton motive force, *Biochim. Biophys. Acta*, 1847, pp. 931–938.

Wang, X., Cao, J., Maroti, P., Stilz, H. U., Finkele, U., Lauterwasse, C., Zinth, W., Oesterhelt, D., Govindjee and Wraight, C. A. (1992). Is bicarbonate in photosystem II the equivalent of the glutamate ligand to the iron atom in bacterial reaction centers? *Biochim. Biophys. Acta*, 1100, pp. 1–8.

Warburg, O. (1964). Prefactory chapter, *Annu. Rev. Biochem.*, 33, pp. 1–18.

Warburg, O. and Krippahl, G. (1958). Hill-Reaktionen, *Z. Naturforsch. B*, 13, pp. 509–514.

Wei, X., Su, X., Cao, P., Liu, X., Chang, W., Li, M., Zhang, X. and Liu, Z. (2016). Structure of spinach photosystem II–LHCII supercomplex at 3.2 Å resolution, *Nature*, 534, pp. 69–74.

Wydrzynski, T. and Govindjee. (1975). New site of bicarbonate effect in photosystem II of photosynthesis – Evidence from chlorophyll fluorescence transients in spinach-chloroplasts, *Biochim. Biophys. Acta,* 387, pp. 403–408.

Van Rensen, J. J. S. and Klimov, V. V. (2005). Bicarbonate interactions. *In* Wydrzynski, T., Satoh, K., eds, *Photosystem II. The Light-Driven Water: Plastoquinone Oxidoreductase*, Vol. 22 (Springer, Dordrecht), pp 329–346.

Van Rensen, J. J. S., Xu, C. and Govindjee. (1999) Role of bicarbonate in photosystem II, the water-plastoquinone oxido-reductase of plant photosynthesis, *Physiol. Plantarum*, 105, pp. 585–592.

Xiong, J., Subramaniam, S. and Govindjee. (1998). A knowledge-based three dimensional model of the Photosystem II reaction center of *Chlamydomonas reinhardtii*, *Photosynth. Res.*, 56, pp. 229–254.

Yamori, W., Shikanai, T. and Makino, A. (2015). Photosystem I cyclic electron flow via chloroplast NADH dehydrogenase-like complex performs a physiological role for photosynthesis at low light, *Sci. Rep.*, 5, p. 13908.

Yano, J., Kern, J., Yachandra, V. K., Nilsson, H., Koroidov, S. and Messinger, J. (2015). Light-dependent production of dioxygen in photosynthesis, *Met. Ions Life Sci.*, 15, pp. 13–43.

Zaraiskaya, T., Vassiliev, S. and Bruce, D. (2014). Discovering oxygen channel topology in photosystem II using implicit ligand sampling and wavefront propagation, *J. Comput. Sci.*, 5, pp. 549–555.

Chapter 4

Basics of Photosynthesis: The Carbon Reactions

4.1 C3 Photosynthesis: The Calvin-Benson Cycle

In Chapter 3, we dealt with the basics of photosynthesis, the photosynthetic apparatus, as well as what is usually referred to as "*the light reactions*," although only the first step, the excitation of a pigment molecule, is a true light reaction. It takes place in PSI and in PSII, located in the thylakoid membrane. We now turn to what goes on in the rest of the chloroplast, the stroma (or in the cytoplasm, in case of cyanobacteria). Here, the most important process is the assimilation of carbon dioxide (CO_2), which leads to the formation of carbohydrates. The collective name for these reactions is "*the carbon reactions*." We summarize the relation between the so-called light reactions and the carbon reactions in Fig. 4.1.

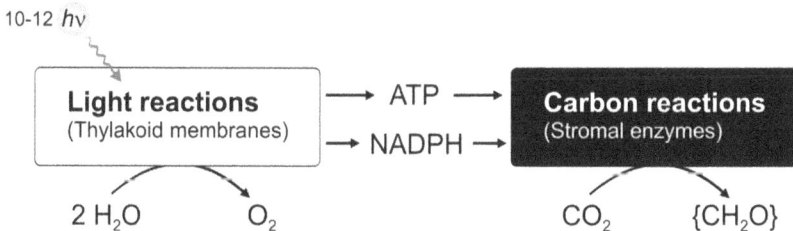

Fig. 4.1. A simplified scheme for the relation between the "light reactions" and the carbon reactions. The end products of the "light reactions" ATP and NADPH provide the energy and the reducing power for the formation of {CH_2O}, a simplified way to express a carbohydrate unit; we emphasize that it does not stand for formaldehyde. Note, that not all reactants and products are shown in this simplified figure. Modified from Buchanan [2016].

Photosynthesis: Solar Energy for Life by Dmitry Shevela, Lars Olof Björn and Govindjee
© 2018, published by World Scientific Publishing Co. Pte. Ltd. ISBN: 978-981-3223-10-3.

Among bacteria there are a number of different variants of carbon reactions, i.e., different ways of carrying out assimilation of CO_2. In addition, there are also other pathways of using ATP and NADPH (or the related NADH), in bacteria, as well as in algae and in plants, which involve assimilation of various forms of inorganic nitrogen. In plants we have three main pathways for the assimilation of CO_2, reflecting adaptation or acclimation to various environmental conditions: (i) C3; (ii) C4; (iii) Crassulacean Acid Metabolism (CAM). The C3 pathway is the Calvin-Benson cycle, but the other two also use it in addition to other pathways; the C4 pathway is called the Hatch and Slack Pathway [Berkowitz *et al.*, 2007]. Figure 4.2 shows a simplified form of the Calvin-Benson cycle [Bassham, 2005; Benson, 2005]. The net result of the C3 cycle is as follows:

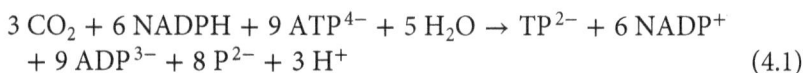

$$3\,CO_2 + 6\,NADPH + 9\,ATP^{4-} + 5\,H_2O \rightarrow TP^{2-} + 6\,NADP^+$$
$$+\,9\,ADP^{3-} + 8\,P^{2-} + 3\,H^+ \tag{4.1}$$

where P stands for phosphate, and TP stands for triose phosphate.

The regeneration of ribulose bisphosphate (RuBP) from TP (e.g., glyceraldehyde 3-phosphate) proceeds *via* a number of reactions involving several sugar phosphates (Fig. 4.2). We can get an overview of the process by simply looking at the number of carbon atoms (C), and the phosphate groups (P) involved:

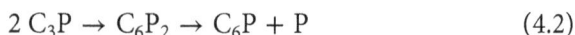

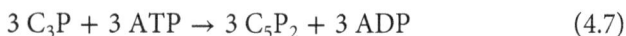

$$2\,C_3P \rightarrow C_6P_2 \rightarrow C_6P + P \tag{4.2}$$

$$C_6P + C_3P \rightarrow C_4P + C_5P \tag{4.3}$$

$$C_4P + C_3P \rightarrow C_7P_2 \tag{4.4}$$

$$C_7P_2 \rightarrow C_7P + P \tag{4.5}$$

$$C_7P + C_3P \rightarrow 2\,C_5P \tag{4.6}$$

$$3\,C_3P + 3\,ATP \rightarrow 3\,C_5P_2 + 3\,ADP \tag{4.7}$$

When we follow it up with the carboxylation, phosphorylation, and reduction reactions, we have the following (PG stands for phosphoglycerate):

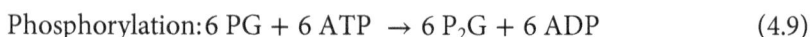

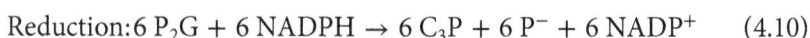

Carboxylation: $3\,C_5P_2 + 3\,CO_2 \rightarrow 6\,PG$ $\hspace{2cm}$ (4.8)

Phosphorylation: $6\,PG + 6\,ATP \rightarrow 6\,P_2G + 6\,ADP$ $\hspace{1cm}$ (4.9)

Reduction: $6\,P_2G + 6\,NADPH \rightarrow 6\,C_3P + 6\,P^- + 6\,NADP^+$ $\hspace{0.3cm}$ (4.10)

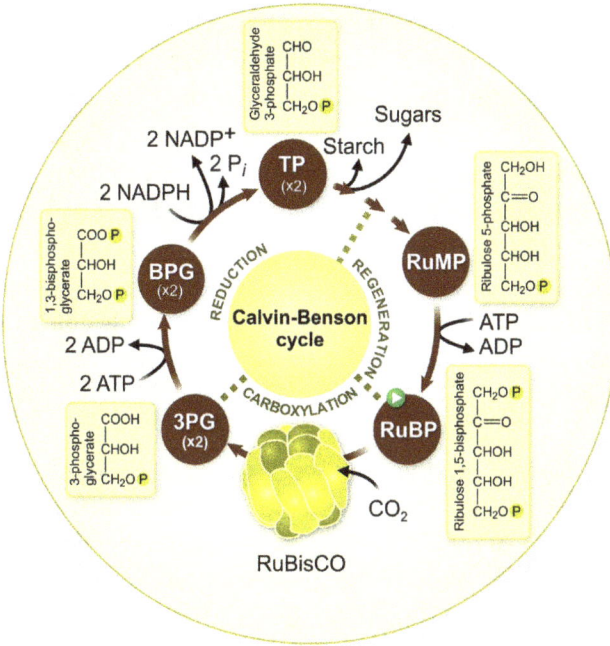

Fig. 4.2. The simplified Calvin-Benson cycle (the C3 cycle). RuBP stands for ribulose 1,5-bisphosphate and 3PG for 3-phosphoglycerate. The 3PG is phosphorylated by ATP to BPG (1,3-biphosphoglycerate), which is reduced to TP (triose phosphate, e.g., glyceraldehyde 3-phosphate) by NADPH with the release of orthophosphate. There are two interconvertible forms of TP, i.e., glyceraldehyde phosphate and dihydroxyacetone phosphate. Further, a fraction of TP molecules leave the cycle to form starch or just go out of the chloroplast. Another intermediate in the Calvin-Benson cycle that participates in the regeneration of RuBP is ribulose 5-phosphate (RuMP). Modified from poster "Oxygenic Photosynthesis" by Shevela and Govindjee [2016].

We note that in the cycle (Fig. 4.3) 5 C_3P (triose phosphates) are consumed and 6 C_3P are produced, the extra one is produced from the three molecules of CO_2.

The above is a simplified overview. A complexity of the process is that, just as there are two kinds of TPs, there are two kinds of pentose-phosphate (C_5P), i.e., ribose phosphate and ribulose phosphate, in addition to RuBP. The entire process also includes a C_4 sugar (erythrose), and a C_7 sugar (seduheptulose). All these reactions are catalyzed

61

by different enzymes [Benson 2005]. Many of the reactions are reversible, but the enzymes are strictly regulated and active only under reducing conditions, when their sulfhydryl groups are in reduced state (—SH), not in oxidized state (—SS—). Thus, the cycle cannot (and does not) run backwards when NADPH is not available.

The carboxylation of one molecule of RuBP yields two molecules of 3-phosphoglycerate, or 3-phosphoglyceric acid (Fig. 4.2). This reaction is catalyzed by *ribulose-bisphosphate carboxylase/oxygenase*, referred to as *RuBisCO*, which is known to have interesting and unique properties [Spreitzer and Salvucci, 2002]. This enzyme is not only able to carboxylate (i.e., add CO_2) RuBP, but also oxygenate it as (also see Fig. 4.3):

$$C_5P_2 + O_2 \rightarrow \text{phosphoglycolate}^{2-} + PG + 2H^+ \qquad (4.11)$$

The phosphoglycolate (the 2-carbon product in Fig. 4.3) is hydrolyzed to glycolate and orthophosphate, and the glycolate is then exported from the chloroplast to another organelle, a peroxisome, where another O_2 molecule is taken up. After further processing in the peroxisome as well as in a mitochondrion and the release of a molecule of CO_2, half of the carbon is salvaged for reuse in the Calvin-Benson cycle. This process is known as *photorespiration*, which decreases the incorporation of carbon from CO_2 into organic material, and is, thus, wasteful under many conditions [Ogren, 2003]. The reason why photorespiration exists is not entirely clear. Part of the explanation is probably that the enzyme RuBisCO evolved

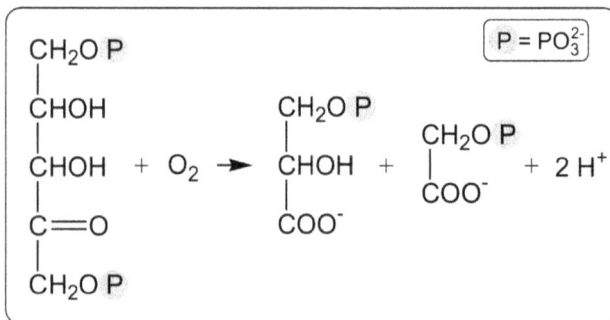

Fig. 4.3. Oxygenation of RuBP by RuBisCO.

under conditions when molecular O_2 was almost completely absent in the environment, while the concentration of CO_2 was higher than at present, and therefore photorespiration did not take place.

Another remarkable property of RuBisCO is that it is extremely sluggish, and therefore a lot of it is required for the carboxylation to proceed at a rate comparable to the other reactions of the cycle. It is one of the most abundant proteins on Earth [Wildman, 2002].

It is surprising that RuBisCO has not become any better than what it was 3 billion years ago; it still does not have high affinity for CO_2, it still does not discriminate well between CO_2 and O_2, and it still is slow in its catalytic action. However, it is interesting to note that Savir *et al.* [2010] and Shih *et al.* [2016] have found that RuBisCO is almost as good as is physically possible! It seems that there is a trade-off between CO_2 specificity and maximum carboxylation rate (Fig. 4.4).

Since molecular oxygen competes with CO_2 for the catalytic site on RuBisCO, the wasteful photorespiration would be decreased if the concentration of CO_2 at the enzyme surface could somehow be increased. Cyanobacteria and algae have ways of doing this by pumping CO_2 or bicarbonate ions across membranes; further, they use carbonic anhydrase for efficient conversion of CO_2 to bicarbonate, and *vice versa*. Thus, it seems worth hoping that transgenic plants will be produced, in the future, by transferring genes from cyanobacteria to higher plants to increase CO_2 fixation, and, thus, biomass! Some plants have evolved another method, which is described below.

4.2 C4 Photosynthesis: The Hatch–Slack Pathway

The C4 photosynthetic carbon assimilation cycle (or C4 pathway), discovered by Hatch and Slack [1966], is best known from angiosperms (both from more than 5000 monocot species and from about 1000 dicot species), but it also occurs in some other plants as well as in some algae [Keeley, 1999; Xu *et al.*, 2012]. For a comprehensive overview, see Raghavendra and Sage [2011]. For a historical perspective, see Hatch [2002]. In most cases its two main phases are divided between two kinds of cells, but this is not always the case [von Caemmerer *et al.*, 2014]. Typically, the primary

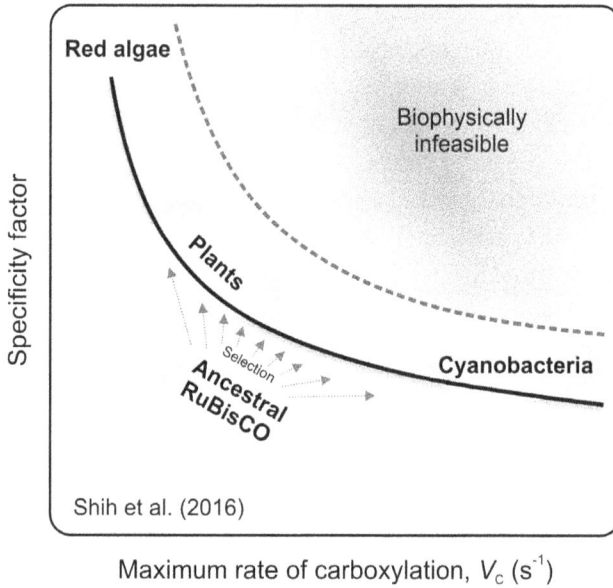

Maximum rate of carboxylation, V_c (s^{-1})

Fig. 4.4. Specificity factor versus rate of carboxylation (V_C) for the characterized RuBisCOs, measured as the ratio of the catalytic efficiency of carboxylation to oxygenation (VoKc)/(VcKo). Here Vc and Vo represent the maximum rates for carbon dioxide and oxygen uptake, respectively, and Kc and Ko the corresponding Michaelis–Menten constants. The best-fit for all the existing RuBisCOs is shown by a black solid line. Model of selective pressures pushing properties of RuBisCO towards the hypothesized upper limits of the kinetic parameters is represented by this best-fit curve. Modified from Shih *et al.* [2016].

binding of CO_2 takes place in *mesophyll cells* away from the vascular tissue in a leaf, while the final processing of CO_2 occurs in *bundle sheath cells* that are around vascular bundles. In many cases, but not all, the C4 pathway is associated with the so-called *Kranz anatomy* (Fig. 4.5). Leaves with typical Kranz anatomy have large cells surrounding the vascular bundles packed with chloroplasts [Sage *et al.*, 2014].

The C4 pathway has evolved independently at least 60 times in various plant groups [Sage *et al.*, 2011; Reyna-Llorens *et al.*, 2018]; thus, it is not surprising that the details differ among plants. What is common in all cases is that the primary binding is not of CO_2, but of bicarbonate ion, and that the binding is to phosphoenolpyruvate (PEP) *via* the enzyme phosphoenolpyruvate carboxylase (PEPC). In plants with Kranz anatomy,

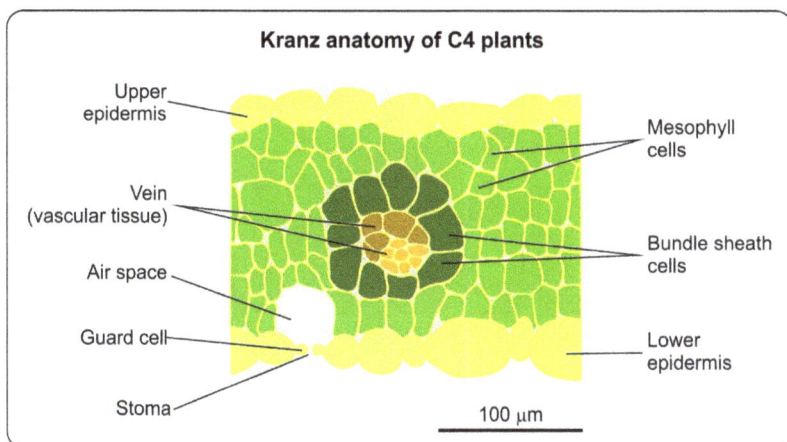

Fig. 4.5. Schematic representation of a transverse section of a leaf with Kranz anatomy (C4 photosynthesis).

this reaction takes place in mesophyll cells. Further, the carbon is translocated in an organic form to bundle sheath cells, where CO_2 is released, and a second processing of CO_2 takes place there *via* the Calvin-Benson cycle.

The three main variants of C4 pathway are (i) the NADP-malic enzyme type, (ii) the NAD-malic enzyme type, and (iii) the PEP carboxykinase type (Fig. 4.6). For an overview of all the three types of C4 plants, see Edwards and Voznesenskaya [2011].

The C4 cycle acts as a pump that concentrates CO_2 at the surface of the RuBisCO, which mediates its final binding. This pump action is not without metabolic cost: it requires two extra high-energy phosphate bonds, resulting in a requirement of five ATP molecules per carbon atom assimilated, instead of three as in C3 pathway.

Molecular oxygen does not compete with bicarbonate ions in the reaction catalyzed by PEPC, and this enzyme has a much higher affinity to bicarbonate than RuBisCO has to CO_2. This results in very low photorespiration in all C4 plants.

The C4 photosynthetic carbon fixation evolved mainly during a time when the CO_2 concentration in the atmosphere had dropped, and the climate had become dryer (Fig. 4.7). C4 plants are much more tolerant to drought than the C3 plants are; this is due to the fact that they are able to

65

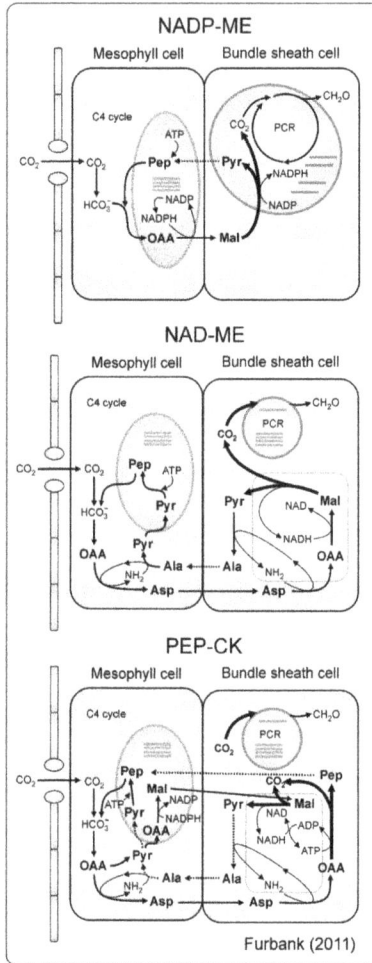

Fig. 4.6. Three variants of the C4 pathway: the NADP-malic enzyme (NADP-ME) type, the NAD-malic enzyme (NAD-ME) type, and the phosphoenolpyruvate carboxykinase (PEP-CK) type. The NADP-ME type differs from the others by lacking well-developed grana and in non-photochemical re-reduction of NADP in the bundle-sheath chloroplasts. Abbreviations: Asp, aspatate; Ala, alanine; Mal, malate; OAA, oxaloacetate; PCR, photosynthetic carbon reduction cycle; Pyr, pyruvate. Chloroplasts and mitochondria are shown by grey circles/ellipses and by white rectangles, respectively. The decarboxylation reactions are highlighted by solid bold arrows. Modified from Furbank [2011].

decrease the aperture of their stomata, and are thus able to conserve water, and yet allowing sufficient influx of CO_2. This is because of the C4 pathway's high affinity for CO_2. They keep a low internal concentration of CO_2

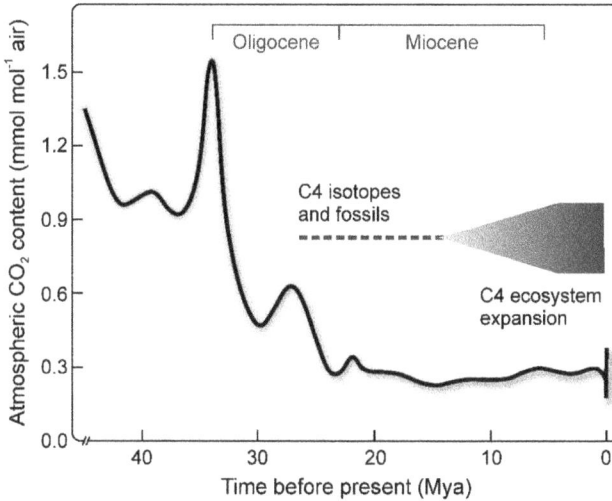

Fig. 4.7. The evolution of C4 pathway coincides with a decrease in atmospheric CO_2 content during Oligocene and Miocene epochs. Modified from Sage *et al.* [2012].

in their intercellular spaces, and thus they maintain a high concentration gradient from that in the external air, making efficient diffusion possible.

The C4 plants are also more tolerant to high temperature than the C3 plants because of their low photorespiration, which has a high temperature coefficient. At the present, most of the C4 plants are largely confined to warm and dry climates, i.e., in the tropics, where they have a competitive advantage over the C3 plants. However, C4 plants are not competitive in cold and wet climates, since they use a greater amount of ATP (see Section 4.2). Surprisingly, the maximum quantum yield of photosynthesis (which is measured on the basis of absorbed photons, in the linear part of light intensity curves) is sometimes higher in the C4 plants than in the C3 plants [Ehleringer and Pearcy, 1983; Ye *et al.*, 2013] (Fig. 4.8).

The chloroplasts in the mesophyll cells of C4 plants differ in several respects from those in the bundle sheath cells; they lack the enzymes of the Calvin-Benson cycle. In the NADP-malic enzyme type, depicted in Fig. 4.6, they have no grana, and at least in some cases they even lack PSII [Woo *et al.*, 1970].

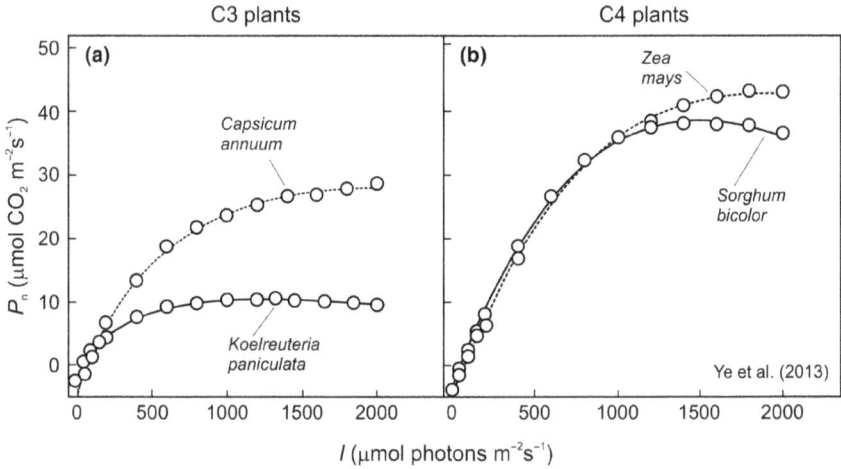

Fig. 4.8. CO_2 uptake as a function of irradiance ("light intensity") for two C3 plants (*Capsicum annuum* and *Koelreuteria paniculata*) and two C4 plants (*Zea mays* and *Sorghum bicolor*). Modified from Ye *et al.* [2013].

We do not discuss here all the variations of C4 pathway, but we do point out one fact—often not emphasized in textbooks—that Kranz anatomy is not necessary for C4 photosynthetic carbon cycle, and in some cases the complete set of reactions is confined to one type of cells, although possessing two separate sets of chloroplasts. This is the so-called single-cell C4 photosynthesis [Voznesenskaya *et al.*, 2002; Edwards *et al.*, 2004; von Caemmerer *et al.*, 2014].

One can estimate that only a couple of percent of plant species are C4 plants, but they account for about 18% of the *net primary productivity* (NPP) on earth [Ehleringer *et al.*, 1997]. The NPP is the *gross primary productivity* (GPP) minus respiration, except photorespiration. For a discussion with definitions of related concepts, see Wohlfahrt [2015].

There are several ways to tentatively determine whether a plant has a C3 or a C4 pathway without detailed biochemical analysis. If several species of plants are enclosed in a glass chamber and illuminated, those which survive longest are likely to have C4 pathway since they are able to assimilate CO_2 until the concentration is almost zero, while C3 plants start giving off CO_2 (respiration exceeding assimilation) when the concentration

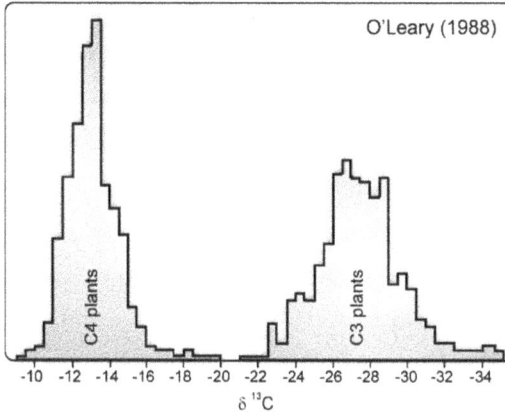

Fig. 4.9. Frequency histograms of $\delta^{13}C$ values (‰) based on ~1000 analyses of plants. The group on the left is for C4 plants, and the group on the right is for C3 plants. Modified from O'Leary [1988].

drops below $\approx$ 70 ppm [Verduin, 1954]. Another way is to determine the $^{13}C/^{12}C$ isotopic composition. The $\delta^{13}C$ values (a measure of the $^{13}C/^{12}C$ ratio, expressed in ‰) of C3 plants are ~ -28‰, whereas that for C4 plants, they are ~ -14‰ (Fig. 4.9). The $\delta^{13}C$ is determined as:

$$\delta^{13}C = 1000 \times \{ [^{13}CO_2/^{12}CO_2(\text{sample})]/ [^{13}CO_2/^{12}CO_2 (\text{standard limestone})]^{-1} \} \quad (4.12)$$

Since the $\delta^{13}C$ has remained unchanged during fossilization, it is possible to even determine the assimilation system for the fossil plants, and, by recovering $\delta^{13}C$ from the teeth of fossil animals, to determine the kind of plants eaten by them or from mummified old Egyptians [Touzeau *et al.*, 2014]. However, care must be exercised in our interpretation since the isotopic discrimination is also influenced by environmental conditions during plant growth [Cernusak *et al.*, 2013].

4.3   Crassulacean Acid Metabolism

The Crassulacean Acid Metabolism (CAM) and C4 pathway have one thing in common: in both the primary binding of CO_2 takes place on PEP and is catalyzed by PEPC; further, there is a secondary fixation and assimilation of CO_2 *via* the Calvin-Benson cycle. But instead of separating the two phases in space (most often in mesophyll and bundle sheath cells, respectively), CAM plants separate the phases temporally (Fig. 4.10). CO_2 is absorbed through open stomata during the night and stored in a carboxy group of malic acid without being reduced. During the day the stomata are closed, CO_2 is released by decarboxylation of the malic acid, and the final assimilation of the CO_2 is carried out through the Calvin-Benson cycle. CAM has the advantage of high water use

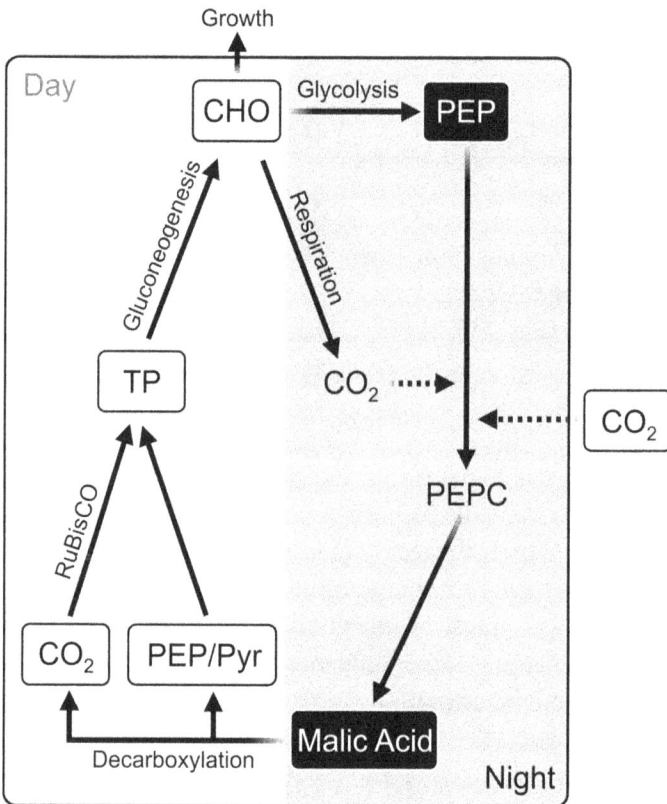

Fig. 4.10. Overview of Crassulacean Acid Metabolism (CAM). CO_2 is taken in at night and stored as carboxy groups in malic acid without being reduced. During the day the CO_2 is reduced and assimilated *via* the Calvin-Benson cycle. Abbreviations: CHO (appropriately, CH_2O), carbohydrate; PEP, phosphoenolpyruvate; PEPC, phosphoenolpyruvate carboxylase; Pyr, pyruvate; TP, triose phosphate. Modified from Borland [2000].

efficiency (high ratio of carbon assimilated to water lost) since stomata are kept closed during the time the plant is exposed to sunlight. CAM plants are thus adapted to very dry environments. The disadvantage of CAM is that they assimilate CO_2 too slowly and, thus, they are not competitive in terrestrial environments with good water supply. However, CAM occurs in many aquatic plants. For historical perspective of CAM discoveries, see a review by Black and Osmond [2003].

4.4   Translocation and Biomass Production

The final product of the Calvin-Benson cycle is TP (the two forms—dihydroxyacetone phosphate and glyceraldehyde phosphate—are readily interconvertible by triose phosphate isomerase). TP is used either for conversion, *via* intermediates, to starch, stored in the chloroplast, or to sucrose or other sugars or organic acids stored in the vacuole, or exported out. This export is *via* the triose phosphate/phosphate translocator in the inner chloroplast membrane; TP is exchanged with orthophosphate. However, in plants accumulating starch, a glucose transporter and a maltose transporter are also essential [Cho *et al.*, 2011]. Furthermore, a large number of other transport proteins are located in the inner chloroplast envelope, while the outer envelope is permeable to most substances with molecular masses up to 10 kDa, in some cases *via* specific transporters (Fig. 4.11) [Flügge, 2000; Breuers *et al.*, 2011; Endow *et al.*, 2016].

In most plants, sucrose is the preferred compound for long-distance transport of carbon, and, thus, the TP, glucose, and maltose, exported from the chloroplast to the mesophyll cytosol, are converted to sucrose (Fig. 4.12). Among the long-distance transported carbon assimilates, sugar alcohols are also important, as well as the so-called raffinose family oligosaccharides, especially the trisaccharide raffinose, the tetrasaccharide stachyose, and the pentasaccharide verbascose.

For transport out of a leaf, assimilates must be transferred from the cytoplasm of the mesophyll cells to the sieve cells or the sieve tubes of the phloem. Several ways of doing this are discussed by several authors [Turgeon and Ayre, 2005; Rennie and Turgeon, 2009; Batashev *et al.*, 2013].

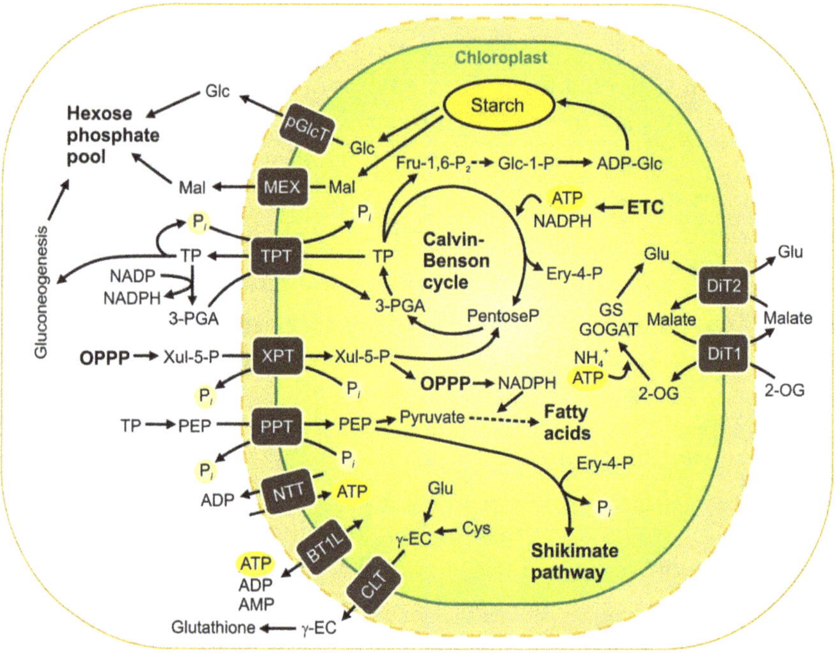

Fig. 4.11. Transport proteins in the inner chloroplast envelope. Those referred to in the text are: the triose phosphate (TP) transporter (TPT), the plastidic glucose (Glc) transporter (pGlcT), and the maltose (Mal) exporter (MEX). Missing from this figure is the vacuole, in which assimilates can be stored as various kinds of sugar. Abbreviations: 2-OG, 2-oxoglutarate; ADP-Glc, ADP-glucose; BT1L, BT1-like transporter; CLT, CL-like transporter; Cys, cysteine; Ery-4-P, erythrose-4-phosphate; ETC, electron transport chain; Fru-1,6-P_2, fructose-1,6-bis-phosphate; γ-EC, γ-glutamylcysteine; Glc-1-P, glucose-1-phosphate; Glu, glutamine; NTT, nucleoside triphosphate transporter; OPPP, oxidative pentose phosphate pathway; 3-PGA (also abbreviated as 3PG; see Fig. 4.2), 3-phosphoglyceric acid; PEP, phosphoenolpyruvate; pGlcT, plastidic glucose transporter; P_i, inorganic phosphate; PPT, PEP phosphate translocator; XPT, xylulose-5-phosphate/phosphate translocator; Xu-5-P, xylulose-5-phosphate. Modified from Facchinelli and Weber, 2011].

In some plants, the first step is the transfer of the assimilate to the apoplast, driven by plasma-membrane transporters and energized by the proton motive force [Sauer, 2007]. This is exemplified by sweet sorghum [Bihmidine *et al.*, 2015].

In other species, loading is symplastic, driven through plasmodesmata from the mesophyll to the phloem by a concentration gradient.

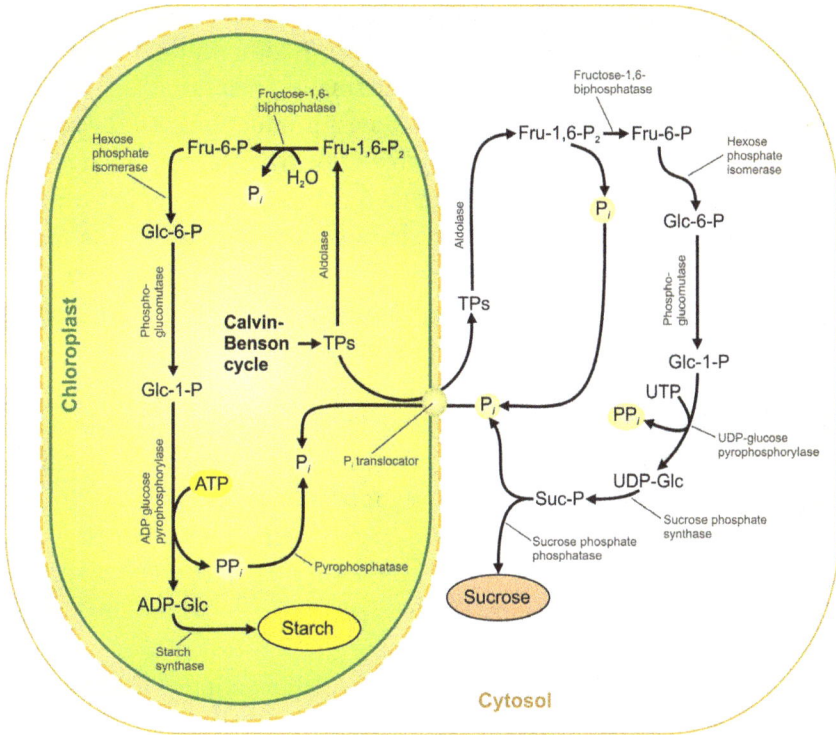

Fig. 4.12. Synthesis of starch and sucrose. Abbreviations: Fru-6-P, fructose-6-phosphate; Glc-6-P, glucose-6-phosphate; PP_i, pyrophosphate; Suc-P, sucrose phosphate; UDP-Glc, uridine diphosphate glucose; UTP, uridine triphosphate; for all other abbreviations, see the legend of Fig. 4.11. As an example of the role and regulation of enzymes shown in the diagram, see Huber and Huber [1996], for sucrose phosphate synthase.

An example of this process is in poplar [Zhang *et al.*, 2014]; further, there are several ways by which this transport takes place.

Inhibition of the transport of the assimilates from the mesophyll cells results in a negative feedback and downregulation of CO_2 assimilation. In Chapter 5, we will describe some of the mechanisms of regulation of photosynthesis.

We emphasize here that photosynthesis should not be confused with biomass production. Not only do all plants consume part of the assimilates by respiration, but the part of assimilates that can be allocated to growth and biomass production varies widely in different plants. Although the mechanism for photosynthesis is very similar in the flowering plants with C3 photosynthesis, the smallest plants have among the highest relative rates of growth and biomass production, and the largest ones the lowest. A strain of *Wolffia globosa* (which is a very small plant) has a relative growth rate (RGR) of 0.56 day^{-1} and a doubling time of 1.24 days [Sree *et al.*, 2015]. However, one of the largest plants *Eucalyptus regnans* is estimated to have an above ground biomass of 215 Mg (megagram = metric tons) and a growth rate of 0.784 Mg year^{-1}, corresponding to an RGR of only 10^{-5} day^{-1} [Sillett *et al.*, 2015]. The larger a plant grows, the more of the assimilate must be allocated to structural support and to long-distance transport. As an example, for a pine tree, 70% of the fixed carbon is respired, 17% allocated to growth, and the rest lost with litter and/or exported to mycorrhiza (a kind of transport investment) and soil [Klein and Hoch, 2015]. Further, and contrary to what is usually believed, some very large trees can maintain and even increase their biomass production per individual over time [Sillett *et al.*, 2010; Sheil *et al.*, 2017].

Globally, the NPP (photosynthetic assimilation minus algal or plant respiration) is 55 Pg (petagram = 10^{15} g) carbon per year in the ocean [Buitenhuis *et al.*, 2013] and 56 Pg C/year on the land [Pan *et al.*, 2015], totaling 111 Pg carbon per year. This corresponds to the amount of CO_2 in the air up to about one and a half km.

4.5 Alternate Pathways and Evolution

There are at least five alternate pathways to the Calvin-Benson cycle, in the Archaea and in the Bacteria. We do not discuss them here, but, for completeness, we refer the readers to Strauss and Fuchs [1993], Fuchs [2011], Björn and Govindjee [2015], and to Buchanan *et al.* [2017], the latter for the Arnon-Buchanan cycle, which is now established in major groups of chemoautotrophic bacteria. For reconstruction of the evolutionary relationships between existing carbon fixation pathways with a single ancestral pathway, see Braakman and Smith [2012].

References

Bassham, J. A. (2005). Mapping the carbon reduction cycle: A personal perspective. *In* Govindjee, Beatty, J. T., Gest, H., Allen, J. F., eds, *Discoveries in Photosynthesis. Advances in Photosynthesis and Respiration* (Springer, Dordrecht), pp. 815–832.

Batashev, D., Pakhomova, M., Razumovskaya, A. and Voitsekhovskaja, O. (2013). Cytology of the minor-vein phloem in 320 species from the subclass Asteridae suggests a high diversity of phloem-loading modes, *Front. Plant Sci.*, 4, p. 312.

Benson, A. A. (2005). Following the path of carbon in photosynthesis: A personal story. *In* Govindjee, Beatty, J. T., Gest, H., Allen, J. F., eds, *Discoveries in Photosynthesis. Advances in Photosynthesis and Respiration* (Springer, Dordrecht), pp. 793–813.

Berkowitz, G. A., Portis, A. R. Jr. and Govindjee. (2007). Carbon dioxide fixation. *In The Encyclopedia of Science and Technology,* 10th ed., Vol. 13 (McGraw Hill Publishers, New York), pp. 475–481.

Bihmidine, S., Baker, R. F., Hoffner, C. and Braun, D. M. (2015). Sucrose accumulation in sweet sorghum stems occurs by apoplasmic phloem unloading and does not involve differential *Sucrose transporter* expression, *BMC Plant Biol.*, 15, p. 186.

Björn, L. O. and Govindjee (2015). The evolution of photosynthesis and its environmental impact. *In* Björn, L. O., ed, *Photobiology: The Science of Light and Life* (Springer, New York), pp. 207–230.

Black, C. C. and Osmond, C. B. (2003). Crassulacean acid metabolism photosynthesis: "Working the night shift", *Photosynth. Res.*, 76, pp. 329–341.

Borland, A. M., Maxwell, K. and Griffiths, H. (2000). Ecophysiology of plants with crassulacean acid metabolism. *In* Leegood, R. C., Sharkey, T. D. and von Caemmerer, S., eds, *Photosynthesis: Physiology and Metabolism* (Kluwer Academic Publishers, Dordrecht), pp. 583–605.

Braakman, R. and Smith, E. (2012). The emergence and early evolution of biological carbon-fixation, *PLoS Comput. Biol.*, 8, p. e1002455.

Breuers, F. K. H., Bräutigam, A. and Weber, A. P. M. (2011). The plastid outer envelope – A highly dynamic interface between plastid and cytoplasm, *Front. Plant Sci.*, 2, p. 97.

Buchanan, B. B. (2016). The carbon (formerly dark) reactions of photosynthesis, *Photosynth. Res.*, 128, pp. 215–217.

Buchanan, B. B., Sirevåg, R., Fuchs, G., Ivanovsky, R. N., Igarashi, Y., Ishii, M., Tabita, F. R. and Berg, I. A. (2017). The Arnon–Buchanan cycle: A retrospective, 1966–2016, *Photosynth. Res.*, 134, pp. 117–131.

Buitenhuis, E. T., Hashioka, T. and Quéré, C. L. (2013). Combined constraints on global ocean primary production using observations and models, *Global Biogeochem. Cycles*, 27, pp. 847–858.

Cernusak, L. A., Ubierna, N., Winter, K., Holtum, J. A. M., Marshall, J. D. and Farquhar, G. D. (2013). Environmental and physiological determinants of carbon isotope discrimination in terrestrial plants, *New Phytol.*, 200, pp. 950–965.

Cho, M.-H., Lim, H., Shin, D. H., Jeon, J.-S., Bhoo, S. H., Park, Y.-I. and Hahn, T.-R. (2011). Role of the plastidic glucose translocator in the export of starch degradation products from the chloroplasts in *Arabidopsis thaliana*, *New Phytol.*, 190, pp. 101–112.

Edwards, G. E., Franceschi, V. R. and Voznesenskaya, E. V. (2004). Single-cell C_4 photosynthesis *versus* the dual-cell (Kranz) paradigm, *Annu. Rev. Plant Biol.*, 55, pp. 173–196.

Edwards, G. E. and Voznesenskaya, E. V. (2011). C4 photosynthesis: Kranz forms and single-cell C4 in terrestrial plants. *In* Raghavendra, A. S., Sage, R. F., eds, *C4 Photosynthesis and Related CO_2 Concentrating Mechanisms* (Springer, Dordrecht), pp. 29–61.

Ehleringer, J. and Pearcy, R. W. (1983). Variation in quantum yield for CO_2 uptake among C_3 and C_4 plants, *Plant Physiol.*, 73, pp. 555–559.

Ehleringer, J. R., Cerling, T. E. and Helliker, B. R. (1997). C4 photosynthesis, atmospheric CO_2, and climate, *Oecologia*, 112, pp. 285–299.

Endow, J. K., Rocha, A. G., Baldwin, A. J, Roston, R. L., Yamaguchi, T., Kamikubo, H. and Inoue, K. (2016). Polyglycine acts as a rejection signal for protein transport at the chloroplast envelope, *PLoS ONE*, 11, p. e0167802.

Facchinelli, F. and Weber, A. (2011). The metabolite transporters of the plastid envelope: An update, *Front. Plant Sci.*, 2, p. 50.

Flügge, U.-I. (2000). Transport in and out of plastids: Does the outer envelope membrane control the flow? *Trends Plant Sci.*, 5, pp. 135–137.

Fuchs, G. (2011). Alternative pathways of carbon dioxide fixation: Insights into the early evolution of life? *Annu. Rev. Microbiol.*, 65, pp. 631–658.

Furbank, R. T. (2011). Evolution of the C4 photosynthetic mechanism: Are there really three C4 acid decarboxylation types? *J. Exp. Bot.*, 62, pp. 3103–3108.

Hatch, M. D. (2002). C4 photosynthesis: Discovery and resolution, *Photosynth. Res.*, 73, pp. 251–256.

Hatch, M. D. and Slack, C. R. (1966). Photosynthesis by sugar-cane leaves: A new carboxylation reaction and the pathway of sugar formation, *Biochem. J.*, 101, pp. 103–111.

Huber, S. C. and Huber J. L. (1996). Role and regukation of sucrose phosphate synthase, *Annu. Rev. Plant Biol. Plant Mol. Biol.*, 47, pp. 431–444.

Keeley, J. E. (1999). Photosynthetic pathway diversity in a seasonal pool community, *Funct. Ecol.*, 13, pp. 106–118.

Klein, T. and Hoch, G. (2015). Tree carbon allocation dynamics determined using a carbon mass balance approach, *New Phytol.*, 205, pp. 147–159.

Ogren, W. L. (2003). Affixing the O to RuBisCO: Discovering the source of photorespiratory glycolate and its regulation, *Photosynth. Res.*, 76, pp. 53–63.

O'Leary, M. H. (1988). Carbon isotopes in photosynthesis: Fractionation techniques may reveal new aspects of carbon dynamics in plants, *BioScience*, 38, pp. 328–336.

Pan, S., Tian, H., Dangal, S. R. S., Ouyang, Z., Lu, C., Yang, J., Tao, B., Ren, W., Banger, K., Yang, Q. and Zhang, B. (2015). Impacts of climate variability and extremes on global net primary production in the first decade of the 21st century, *J. Geogr. Sci.*, 25, pp. 1027–1044.

Raghavendra, A. S. and Sage, R., eds (2011). *C4 Photosynthesis and Concentrating Mechanisms* (Springer, Dordrecht).

Rennie, E. A. and Turgeon, R. (2009). A comprehensive picture of phloem loading strategies, *Proc. Natl. Acad. Sci. U.S.A.*, 106, pp. 14162–14167.

Reyna-Llorens, I., Burgess, S. J., Reeves, G., Singh, P., Stevenson, S. R., Williams, B. P., Stanley, S. and Hibberd, J. M. (2018). Ancient duons may underpin spatial patterning of gene expression in C_4 leaves, *Proc. Natl. Acad. Sci. U.S.A.*, 115, pp. 1931–1936.

Sage, R. F., Christin, P.-A. and Edwards, E. J. (2011). The C4 plant lineages of planet Earth, *J. Exp. Bot.*, 62, pp. 3155–3169.

Sage, R. F., Khoshravesh, R. and Sage, T. L. (2014). From proto-Kranz to C4 Kranz: Building the bridge to C4 photosynthesis, *J. Exp. Bot.*, 65, pp. 3341–3356.

Sage, R. F., Sage, T. L. and Kocacinar, F. (2012). Photorespiration and the evolution of C_4 photosynthesis, *Annu. Rev. Plant Physiol.*, 63, pp. 19–47.

Sauer, N. (2007). Molecular physiology of higher plant sucrose transporters, *FEBS Lett.*, 581, pp. 2309–2317.

Savir, Y., Noor, E., Milo, R. and Tlusty, T. (2010). Cross-species analysis traces adaptation of RuBisCO toward optimality in a low-dimensional landscape, *Proc. Natl. Acad. Sci. U.S.A.*, 107, pp. 3475–3480.

Shevela, D. and Govindjee (2016). Poster "Oxygenic Photosynthesis", unpublished.

Sheil, D., Eastaugh, C. S., Vlam, M., Zuidema, P. A., Groenendijk, P., Sleen, P., Jay, A., Vanclay, J. and Larjavaara, M. (2017). Does biomass growth increase in the largest trees? Flaws, fallacies and alternative analyses. *Funct. Ecol.*, 31, pp. 568–581.

Shih, P. M., Occhialini, A., Cameron, J. C., Andralojc, P. J., Parry, M. A. J. and Kerfeld, C. A. (2016). Biochemical characterization of predicted Precambrian RuBisCO, *Nature Commun.*, 7, p. 10382.

Sillett, S. C., Van Pelt, R., Koch, G. W., Ambrose, A. R., Carroll, A. L., Antoine, M. E. and Mifsud, B. M. (2010). Increasing wood production through old age in tall trees, *For. Ecol. Manage.*, 259, pp. 976–994.

Sillett, S. C., Van Pelt, R., Kramer, R. D., Carroll, A. L. and Koch, G. W. (2015) Biomass and growth potential of *Eucalyptus regnans* up to 100 m tall, *Forest Ecol. Manage.*, 348, pp. 78–91.

Spreitzer, R. J. and Salvucci, M. E. (2002). RuBisCO: Structure, regulatory interactions, and possibilities for a better enzyme, *Annu. Rev. Plant Biol.*, 53, pp. 449–475.

Sree, K. S., Sudakaran, S. and Appenroth, K.-J. (2015). How fast can angiosperms grow? Species and clonal diversity of growth rates in the genus *Wolffia* (Lemnaceae), *Acta Physiol. Plant.*, 37, p. 204.

Strauss, G. and Fuchs, G. (1993). Enzymes of a novel autotrophic CO_2 fixation pathway in the phototrophic bacterium *Chloroflexus aurantiacus*, the 3-hydroxypropionate cycle, *Eur. J. Biochem.*, 215, pp. 633–643.

Touzeau, A., Amiot, R., Blichert-Toft, J., Flandrois, J. P., Fourel, F., Grossi, V., Martineau, F., Richardin, P. and Lecuyer, C. (2014). Diet of ancient Egyptians inferred from stable isotope systematics, *J. Archaeol. Sci.*, 46, pp. 114–124.

Turgeon, R. and Ayre, B. G. (2005). Pathways and mechanisms of phloem loading. *In* Holbrook, N. M., Zwieniecki, M. A., eds, *Vascular Transport in Plants* (Elsevier, Boston), pp. 45–67.

Verduin, J. (1954). Carbon dioxide compensation point in photosynthesis, *Science*, 120, pp. 75–76.

von Caemmerer, S., Edwards, G. E., Koteyeva, N. and Cousins, A. B. (2014). Single cell C4 photosynthesis in aquatic and terrestrial plants: A gas exchange perspective, *Aquat. Bot.*, 118, pp. 71–80.

Voznesenskaya, E. V., Franceschi, V. R., Kiirats, O., Artyusheva, E. G., Freitag, H. and Edwards, G. E. (2002). Proof of C4 photosynthesis without Kranz anatomy in *Bienertia cycloptera* (Chenopodiaceae), *Plant J.*, 31, pp. 649–662.

Wildman, S. G. (2002). Along the trail from Fraction I protein to Rubisco (ribulose bisphosphate carboxylase oxygenase). *Photosynth. Res.*, 73, pp. 243–250.

Wohlfahrt, G. and Gu, L. (2015). The many meanings of gross photosynthesis and their implication for photosynthesis research from leaf to globe, *Plant Cell Environ.*, 38, pp. 2500–2507.

Woo, K. C., Anderson, J. M., Boardman, N. K., Downton, W. J. S., Osmond, C. B. and Thorne, S. W. (1970). Deficient photosystem II in agranal bundle sheath chloroplasts of C4 plants, *Proc. Natl Acad. Sci. U.S.A.*, 67, pp. 18–25.

Xu, J., Fan, X., Zhang, X., Xu, D., Mou, S., Cao, S., Zheng, Z., Miao, J. and Ye, N. (2012). Evidence of coexistence of C3 and C4 photosynthetic pathways in a green-tide-forming alga, *Ulva prolifera*, *PLoS ONE*, 7, p. e37438.

Ye, Z.-P., Suggett, D. J., Robakowski, P. and Kang, H.-J. (2013). A mechanistic model for the photosynthesis–light response based on the photosynthetic electron transport of photosystem II in C3 and C4 species, *New Phytol.*, 199, pp. 110–120.

Zhang, C., Han, L., Slewinski, T. L., Sun, J., Zhang, J., Wang, Z.-Y. and Turgeon, R. (2014). Symplastic phloem loading in poplar, *Plant Physiol.*, 166, p. 306.

Chapter 5

Regulation of Photosynthesis

5.1 Background

The distant nuclear reactor, the Sun, has maintained a remarkably constant output of energy over billions of years; this has allowed the evolution of biosphere and of photosynthesis on our planet. However, the Earth rotates, and, therefore, the light available to us varies over the day and the night. Thus, not only the humans, but also the photosynthetic apparatus must cope with this daily rhythm. There are also variations in light (and temperature) over the year. These variations are relatively slow and regular. However, clouds pass over the sky, and the wind moves the leaves in canopies in an unpredictable manner, causing irregular and often rapid changes in the light falling on a plant. The plant must balance the rate of "electron transport" in its two photosystems to have an optimum maximum reaction, providing the needed amount of the reducing power and the "high energy" phosphate (in ATP). When in high light, plants must also have appropriate "safety valves" for dissipating the excess energy, which, otherwise, would damage the plant. Plants must be able to work not only in high light, but also when the raw materials for the assimilation process become scarce due to the presence of damaging conditions such as drought, or extremely high or very low temperatures. Unfortunately, plants do not have any *social security*; they cannot move to protected areas when conditions get rough and tough. The animals handle the adverse conditions by changes in their behavior, and/or movement to safer areas, but the plants must handle them mainly by changes in their physiology and their metabolism.

Photosynthesis: Solar Energy for Life by Dmitry Shevela, Lars Olof Björn and Govindjee
© 2018, published by World Scientific Publishing Co. Pte. Ltd. ISBN: 978-981-3223-10-3.

Regulation of photosynthesis takes place in many different ways. For a basic background on the *relation of photosynthesis to physiology and ecology*, see Ehleringer [2006], and for specific topics on the *regulation of photosynthesis*, see Aro and Andersson [2011] and Derks *et al.* [2015]. We shall first deal with the regulatory system in the thylakoid membranes. Next, we will discuss the enzymatic reactions in the stroma. Finally, we shall deal with regulation at microscopic and macroscopic levels: movement of chloroplasts and of leaves, and of plant growth.

The state of the electron transport chain is sensed in at least two ways: (i) by lumenal pH, and (ii) by the redox state of the electron transport chain between the two photosystems—often due to imbalance in light absorption in the two photosystems. Both these sensing systems regulate protein phosphorylation, which, in turn, leads to adjustment of light harvesting. We begin our discussion with a specific regulation phenomenon, the *state transitions*, by which plants, algae and cyanobacteria regulate light energy distribution between the two photosystems for maximal efficiency.

5.2 State Transitions

A mechanism is necessary for matching the rates of the two photosystems when light is limiting, since these two have different absorption, and, thus, action spectra, and the spectrum of available light varies over the day. In strong light, there is damage (by photoinhibition) often involving *reactive oxygen species* (ROS); thus, the important thing is to take care of excess energy in a safe way, as described below. Long term matching of the rates of the two photosystems occurs by gene regulation of the stoichiometry, but we shall not discuss this here.

A general consensus is that the short-term matching of the rates of PSI and PSII occurs by the so-called *state transitions* [Murata, 1969, 2009; Bonaventura and Myers, 1969; Duysens, 1972; Allen and Mullineaux, 2004; Minagawa, 2011; Papageorgiou and Govindjee, 2011] (Fig. 5.1). The key point is the sensing of the redox state of the PQ pool and of the Cyt b_6f complex; these components are physically located in between the two photosystems (Fig. 2.5). Cyt b_6f has two binding sites for plastoquinone/ plastoquinol, the Q_o-site (quinol oxidase, on the lumenal side; the "*p*" side)

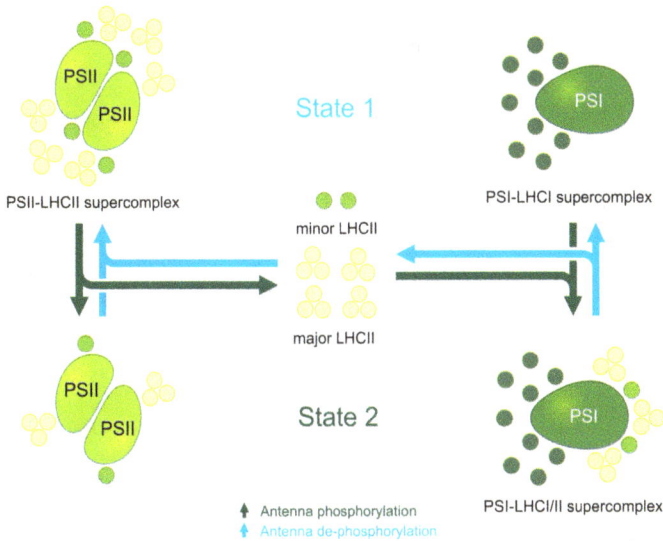

Fig. 5.1. Schematic representation of state transitions. Modified after Goldschmidt-Clermont and Bassi [2015].

and the Q_i-site (quinone reductase, on the stromal side; the "n" side). (Note that the "p" and "n" sides refer to positive and negative sides of the membrane after proton transfer has taken place [Gavel *et al.*, 1991].) Redox sensing takes place at the Q_o-site (Fig. 5.2). According to Zito *et al.* [1999], reduction of the PQ pool, by PSII, and then the binding of a plastoquinol molecule at Q_o, involving a conformational change in the Cyt b_6f, triggers a signal that activates a LHCII kinase. The same activation can also be triggered by acidification of the luminal compartment. The above conclusions were obtained from experiments with the green alga *Chlamydomonas reinhardtii*; they were then confirmed in higher plant chloroplasts [Hou *et al.*, 2002]. In *Chlamydomonas* the kinase is called Stt7 [Depège *et al.*, 2003], but in *Arabidopsis*, it is called STN7 [Bellafiore *et al.*, 2005].

The activated kinase is responsible for phosphorylation of the "mobile" Lhcb1 and Lhcb2, which leads to their detachment from the PSII core and diffusion from the appressed thylakoid region in the grana out into the stroma thylakoid region, and their eventual attachment to PSI. The reason

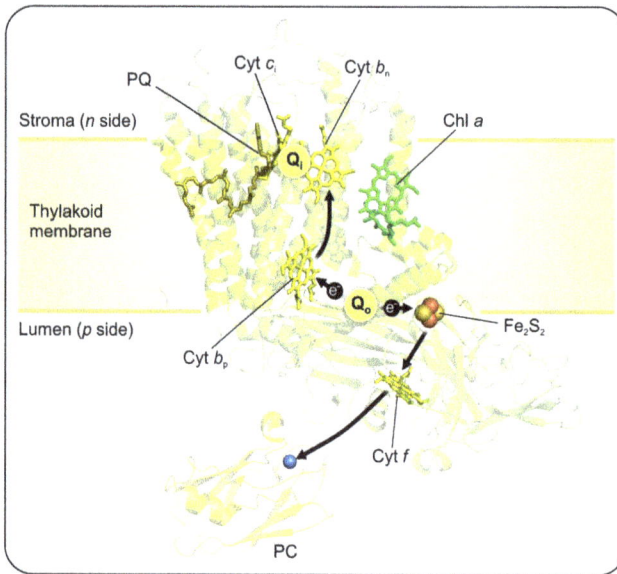

Fig. 5.2. Arrangement of electron-transfer cofactors and Q_o and Q_1 sites in the structure of the Cyt b_6f complex. The arrows indicate the direction of the electron transfer. Also shown is the small luminal water-soluble copper protein, PC, which accepts electrons from the reduced Cyt f in the Cyt b_6f and delivers them to the oxidized form of P700 of PSI (see Chapter 3; Figs. 3.1 and 3.6). The protein background and arrangement of the cofactors of the Cyt b_6f complex are based on the coordinates deposited at PDB with ID 1VF5 [Kurisu *et al.*, 2003]. PC was modeled using the coordinates from PDB with ID 2GIM [Schmidt *et al.*, 2006]. Modified from Tikhonov [2014].

for the detachment is that the negative charges on phosphate repel each other. This situation is called *State 2* since it is initiated by light 2, i.e., light absorbed in pigment system 2. The end result is that the energy input to PSII is diminished, and that to PSI is increased. Since the quantum yield of PSI Chl fluorescence is much lower than that of PSII, the overall fluorescence yield in State 2 is low, as compared to that in State 1, as discussed later. Complete phosphorylation of Lhcb1 and Lhcb2 requires only moderate reduction of the PQ pool [Pursiheimo *et al.*, 2003]. Crepin and Caffari [2015] and Longoni *et al.* [2015] have, however emphasized the role of phosphorylation of Lhcb2, over that of LHcb1 for state transitions.

On the other hand, when the PQ pool is highly oxidized, Lhcb1 and Lhcb2, bound to PSI, are dephosphorylated by a phosphatase, TAP38

[Pribil *et al.*, 2010], also known as PPH1 [Shapiguzov *et al.*, 2010], and the system goes to State 1; it is called *State 1* since it is initiated by light 1, i.e., light absorbed in pigment system 1. Now, the mobile Lhcbs leave the PSI area and return to PSII and the fluorescence yield becomes high, as implied above. Here, there is dephosphorylation of mobile Lhcbs due to phosphorylase activity (constitutively present).

In State 1, when all of LHCIIs are connected to PSII and none to PSI, PSII absorbs more light than PSI (Fig. 5.3) [Laisk *et al.*, 2014]. (A simple and quick method to observe State 2 to State 1 transition is to monitor the slow S to M fluorescence rise in Chl *a* fluorescence transient [Kaňa *et al.*, 2012; Kodru *et al.*, 2015].) The "state change" is advantageous for a plant partly shaded by other plants, since under these conditions there is much more long-wavelength light, which is preferentially absorbed by PSI. If the plant becomes exposed to more direct sunlight, the change to State 2 would give a more balanced energy supply to the photosystems. New knowledge on the detailed structure of the PSII supercomplex in terrestrial plants is expected to provide further insight into the mechanism of light acclimation under terrestrial environment [Su *et al.*, 2017].

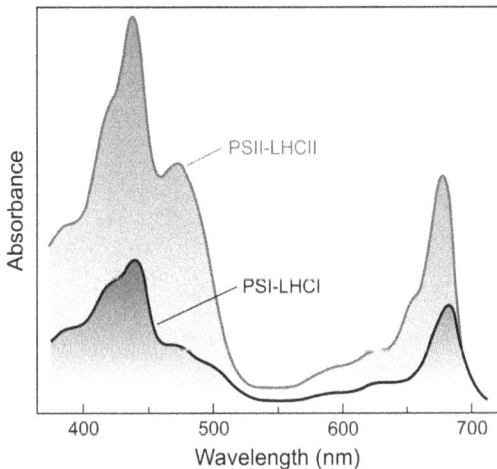

Fig. 5.3. Absorption spectra of PSI and PSII supercomplexes in State 1. Note: In State 2, mobile LHCIIs move to the PSI supercomplex increasing its absorption over that of PSII supercomplex. Modified from Laisk *et al.* [2014].

5.3 Safety Valve: Dissipation of Electronic Energy as Heat ("Quenching")

The state transitions, described above, are important in low light when overall photosynthesis is determined by the "light reactions," and balanced excitation between the two photosystems is essential for optimal electron transport. In high light, photosystems receive sufficient energy, and the overall rate of photosynthesis is determined by the enzymatic reactions. Thus, the important thing is to dispose of the excess energy in a safe way to avoid damage. There are several such "safety valves" involving the pigment antennas, as well as the reaction centers. We know that the antennas of PSI and PSII are not static; they continuously undergo conformational fluctuations. The alternation between fluorescent and more or less quenched states of isolated LHCII complexes is clearly shown by the phenomenon of *blinking* [Chmeliov *et al.*, 2013], as studied by single-molecule spectroscopy [Krüger *et al.*, 2011a, 2011b; Schlau-Cohen *et al.*, 2015; Schörner *et al.*, 2015]. By "blinking" we mean fluctuating fluorescence as shown in Fig. 5.4.

The frequency of different quenched levels of LHCII depends on the pH of the stroma: The efficiency of light harvesting, represented by pigment protein complexes typically residing for long periods in strongly

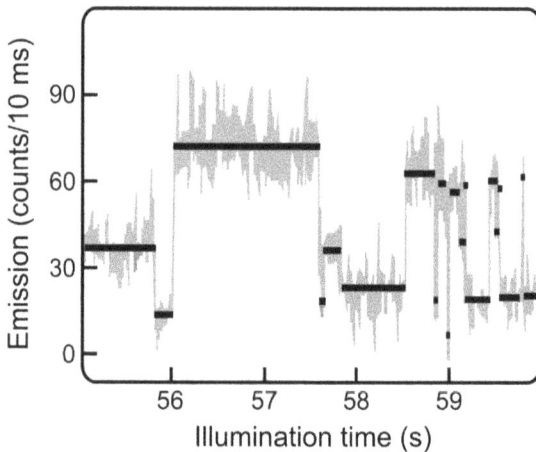

Fig. 5.4. Fluctuations in chlorophyll fluorescence intensity caused by conformational fluctuations in single LHCII trimers. Modified from Krüger *et al.* [2011b].

fluorescing states, decreases when the incident light increases [Krüger *et al.*, 2011b]. Carotenoids (Cars), especially lutein in LHCII, play a major role in excitation quenching; they take the excitation energy and loose it as heat.

The fate of excitation energy is often monitored using Chl *a* fluorescence; the use of energy in photosynthetic reactions is also thought of as "quenching" since it competes with fluorescence. Other processes resulting in decreased fluorescence, as those described in this and the next section, are therefore said to cause *non-photochemical quenching* (NPQ) of the excited state of Chl (see chapters in [Demmig-Adams *et al.*, 2014]; for a timeline of discoveries, see [Papageorgiou and Govindjee, 2014]).

5.4 Xanthophyll Cycles

In plants and algae, there are several kinds of xanthophyll cycles [Latowski *et al.*, 2011; Goss and Lepetit, 2015], of which only the VAZ (violaxanthin-antheraxanthin-zeaxanthin) cycle and the lutein-lutein epoxide cycle are known to exist in higher plants; the VAZ cycle involves transformation between violaxanthin, antheraxanthin, and zeaxanthin (Fig. 5.5).

We know that zeaxanthin is able to quench the excited state of Chl *a* located in the pigment system (Fig. 5.6). But what is the significance of the VAZ cycle? The answer to this is a bit disputed, which is not surprising since the lowest energy level of Cars is difficult to determine. The transition from the ground state to the lowest excited singlet state (S_1) of Cars is "forbidden," which means that it is not visible in the absorption spectrum. However, it can be reached from the ground (S_0) state by energy transfer from another molecule. Although the S_1 level cannot be determined from the absorption spectrum there are several other methods to get that information. A problem is that there is no firm agreement on the interpretation. (For review on the Cars, see chapters in Frank *et al.* [1999].) A further complication is that the transitions between different Cars are associated with structural changes in the thylakoid membranes, which may affect the energy levels.

One of the theories explaining the function of the VAZ cycle (as well as of the diadinoxanthin-diatoxanthin cycle of dinoflagellates) is the "gear

The VAZ Cycle

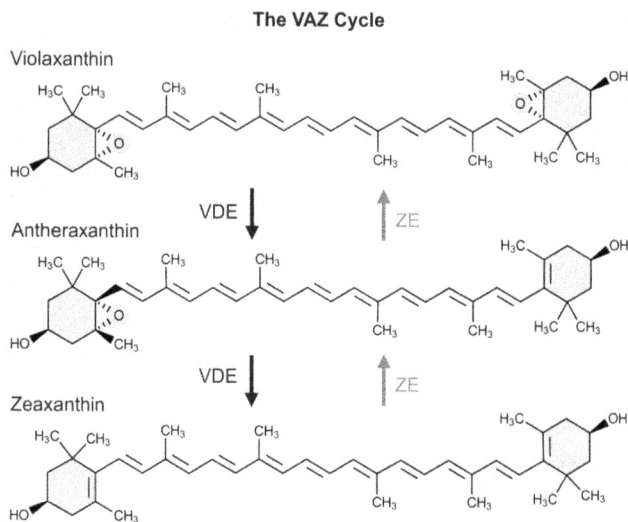

Violaxanthin

VDE ↓   ↑ ZE

Antheraxanthin

VDE ↓   ↑ ZE

Zeaxanthin

Fig. 5.5. The VAZ (violaxanthin-antheraxanthin-zeaxanthin) cycle, also referred to simply as the violaxanthin cycle. Interconversion between the three carotenoids is mediated by violaxanthin de-epoxidase (VDE) and zeaxanthin epoxidase (ZE). Modified from Latowski *et al.* [2011].

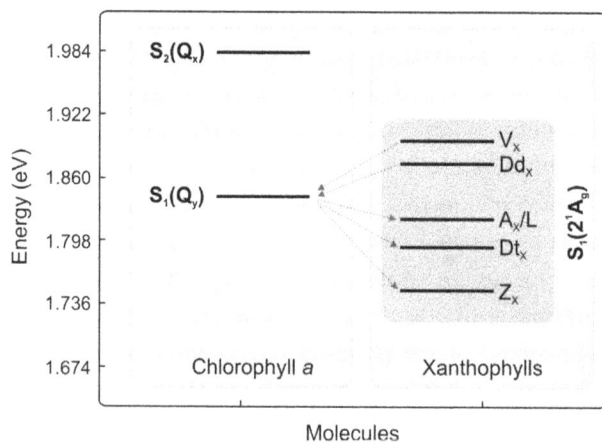

Fig. 5.6. Energy-level diagram for chlorophyll (Chl) *a* (1.984 and 1.822 eV) and various xanthophylls: violaxanthin (Vx, 1.896 eV), diadinoxanthin (Ddx, 1.876 eV), antheraxanthin and lutein (Ax/L, 1.825 eV), diatoxanthin (Dtx, 1.796 eV), and zeaxanthin (Zx, 1.757 eV). Arrows, between the Chl *a* and the xanthophyll box, show energetically favored excitation transfer. Modified from Latowski *et al.* [2011].

shift" mechanism (Fig. 5.6): here the S_1 level of violaxanthin (and diadinoxanthin) is so much higher than the S_1 state of Chl *a* that they function as the antenna pigment and deliver energy to Chl *a*, while the S_1 levels of the other Cars of the xanthophyll cycles are lower than that of Chl *a*, and, thus, they are able to draw excess energy from Chl *a*. However, not everyone agrees on this point. For instance, Polívka *et al.* [1999], Polívka and Sundström [2004], and Frank *et al.* [2000] are of the opinion that the S_1 level of violaxanthin is too low for this mechanism to function. Thus the regulating effect of the VAZ cycle may involve another mechanism: structural (or conformational) change. Changes in light scattering in leaves, probably associated with structural changes in thylakoid membranes, have indeed been observed and correlated with violaxanthin—zeaxanthin transformation. Most research on the violaxanthin cycle has thus far been concerned mainly with photoprotection of PSII (Fig. 5.7)

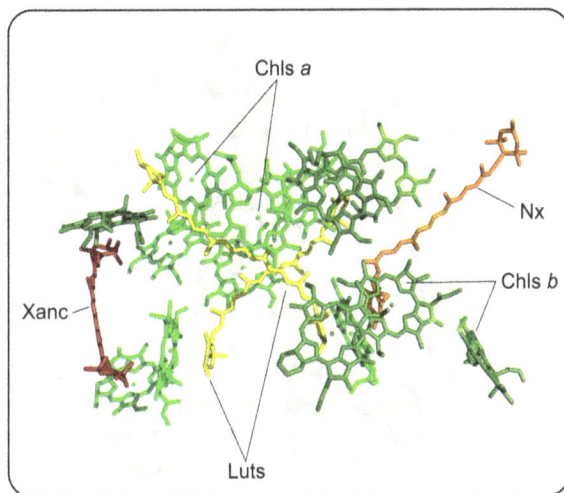

Fig. 5.7. Location of carotenoids and chlorophylls (Chls) in LHCII. The protein backbone with the three transmembrane helices (transparent green) is shown in the background. For Chls, only the tetrapyrrole rings are shown (light green = Chl *a*, and dark green = Chl *b*). Four bound xanthophylls were identified in the structure: two central luteins (Luts, yellow; two binding sites are shown), a xanthophyll cycle pigment (Xanc) on the left (red, V1 site) and neoxanthin (Nx) on the right (orange, N1 site). The LHCII model was generated using x-ray crystallographic coordinates obtained at a resolution of 2.72 Å and deposited at PDB with ID 1RWT. For original x-ray study, see Liu *et al.* [2004].

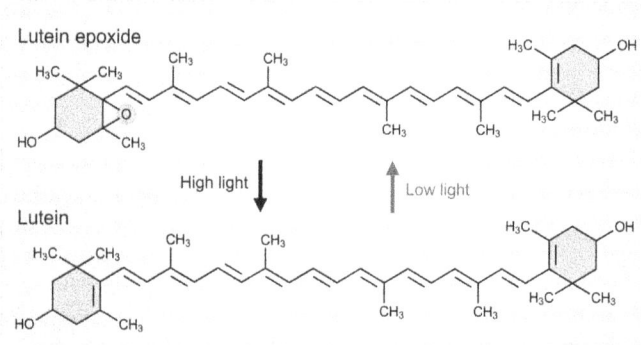

Fig. 5.8. The lutein cycle. Interconversion between two carotenoids, lutein and lutein epoxide, is mediated by violaxanthin de-epoxidase and (probably also) by zeaxanthin epoxidase. Modified from Latowski *et al.* [2011].

[Jahns and Holzwarth, 2012], but it is important to note that it may also be involved in the protection of PSI [Ballottari *et al.*, 2014].

The VAZ cycle appears to be operating in almost all plants and several algae. The other xanthophyll cycle of plants, the lutein cycle (Fig. 5.8), on the other hand, is restricted to some plants, and in particular some woody plants [Garcia-Plazaola *et al.*, 2002; Esteban *et al.*, 2009; Matsubara *et al.*, 2009, 2011]. Here, it is the lutein state that is dissipative (i.e., loss of energy as heat). Further, the lutein cycle operates in parallel with the violaxanthin cycle in avocado leaves [Matsubara *et al.*, 2011].

5.5 Reaction Centers as Sinks for Excess Excitation Energy

In case of excess of excitation energy, electrons from PSII may end up on molecular oxygen, yielding O_2^- (superoxide), which can cause damage, but the plant is equipped with a battery of systems to take care of this. Superoxide dismutase converts superoxide to hydrogen peroxide and water, and catalase converts hydrogen peroxide to water and molecular oxygen. Various antioxidants, such as ascorbate, also help in the protection process. Oxidation of the β-carotene on D2 may be involved in a secondary electron transfer, and it may quench triplet Chl *a*, formed under high light (Fig. 3.7a) [Shinopoulos and Brudvig, 2012]. Further, β-carotene is also able to neutralize destructive singlet oxygen. According to Feyziyev *et al.* [2013] and Ananyev *et al.* [2016] electron flow can go all the way back to the Mn_4CaO_5 cluster in the S_2 and the S_3 states of the OEC (see Fig. 3.7a). Moreover, Ananyev *et al.* [2016], studying phylogenetically diverse species of algae and cyanobacteria, found that light-driven backward electron transitions occur in a large fraction of PSII. Thus, backward transitions were postulated as physiologically important *downregulation* pathways, which contribute to a proton gradient through direct cyclic electron flow within PSII [Ananyev *et al.*, 2016]. Further, photoprotection may occur *via* other, still to be understood, energy-transducing mechanism(s) [Ananyev *et al.*, 2017].

Furthermore, for its regulation PSII uniquely requires bicarbonate ions (see Section 3.3.2.3 in Chapter 3, for the *unique role of bicarbonate in light-induced reactions of PSII*). Brinkert *et al.* [2016] have suggested that, under light, the presence of high concentration of reduced Q_A (Q_A^-) can lead to the release of the "acceptor side" bicarbonate from its binding site at the NHI (Fig. 5.9a,b), and that this downregulates PSII (by making the protonation of Q_B^- [or Q_B^{2-}] less efficient), thereby protecting PSII against photodamage. Additionally, on the "donor" side of PSII, lower levels of dissolved CO_2/HCO_3^- are suggested to reduce the turnover efficiency of the OEC [Shevela *et al.*, 2013], due to the lack of mobile bicarbonate ions that promote water oxidation, by acting as proton acceptors [Shutova *et al.*, 2008; Koroidov *et al.*, 2014]. Thus, under optimal conditions, bicarbonate ions function in electron transfer and protonation events, as described in Section 3.3.2.3 in Chapter 3, allowing PSII to operate at maximum efficiency so that the required number of electrons can be used for CO_2-fixation. However, when the plant is facing drought, high light, or high temperature, stomata may close, resulting in a decrease in the internal concentration of CO_2/HCO_3^- (such a decrease is also expected to occur in algae and cyanobacteria that do not have stomata; [Fukuzawa *et al.*, 2012]). This would lead to a "depletion" of bicarbonate ions from PSII, limiting the activity of the entire PSII enzyme (the water-plastoquinone oxidoreductase; see [Wydrzynski and Satoh, 2005]) leading to a slowdown in CO_2-fixation. As result, it would prevent over-reduction of the PQ-pool

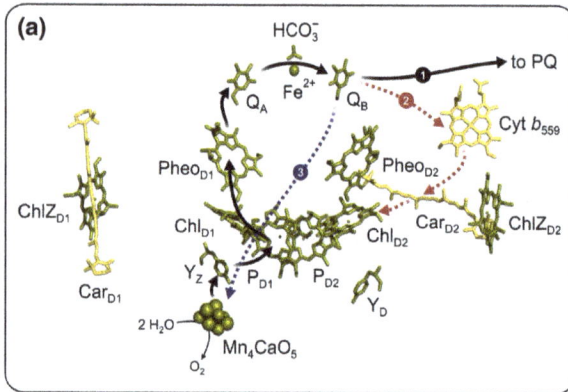

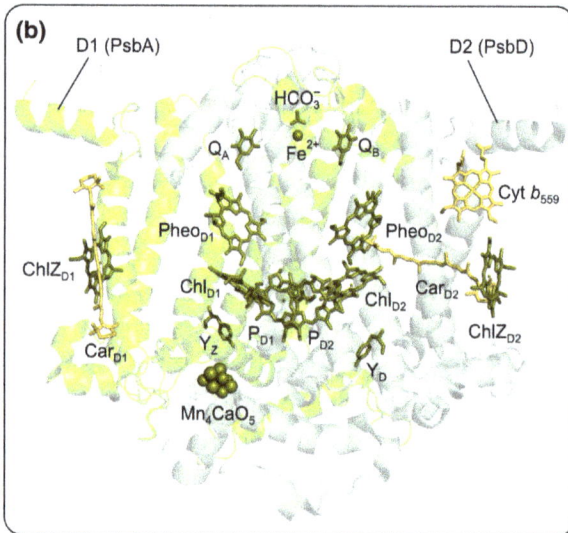

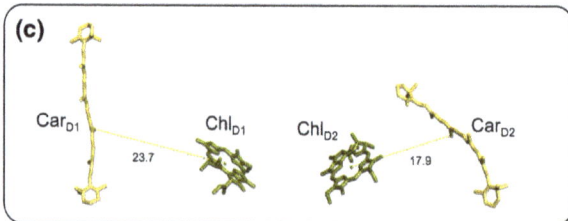

(a)

(b)

D1 (PsbA)   D2 (PsbD)

(c)

(continued)

Fig. 5.9. (continued) Redox cofactors and electron-transfer pathway in PSII. **(a)** Linear and cyclic electron flow in PSII. The linear (normal) path of electrons, to the plastoquinone Q_A and then to Q_B, is *via* Chl_{D1} and $Pheo_{D1}$, and except for the plastoquinone Q_B, the components on the D2 side have not previously been assigned any role (shown by black arrow, labeled as 1). One of the paths of electrons back from the reduced Q_B to the oxidized P_{680} is indicated by dashed dark red arrows (labeled as 2) (according to Shinopoulos and Brudvig [2012]). However, according to Ananyev *et al.* [2016, 2017] the electron flow can go all the way back from the reduced Q_B *via* P680 to the Mn_4CaO_5 cluster (dark dashed arrow, labeled as 3). Other redox cofactors that take part in this PSII-cyclic electron flow have not yet been identified. **(b)** Arrangement of the PSII redox cofactors on D1 and D2 proteins. **(c)** Distances between Car_{D1} and accessory Chl_{D1} pigments of the D1 branch and Car_{D2} and Chl_{D2} of the D2 branch, in the PSII reaction center, are given in angstroms. The arrangement of PSII redox cofactors, protein scaffold, and the distances are based on the data of the x-ray crystallographic PSII structure obtained at 1.9 Å resolution [Umena *et al.*, 2011] and deposited at PDB as ID 3ARC.

and tune the recombination reactions within PSII, thereby reducing the risk of producing harmful ROS and promoting the system to have safe dissipation of excess energy.

We note that both PSII and PSI reaction centers have protective devices built into them [Endo and Asada, 2006; Derks *et al.*, 2015]. Further, in evergreen conifers, during winters, there exists a downregulation mechanism of non-cyclic electron transport that seems to also involve an increase in the cyclic electron transport around PSI [Fréchette *et al.*, 2015].

5.6   Quenching of Chlorophyll Triplets

It is well known that Chl molecules in the singlet state (having electrons spinning in opposite directions) have a distinct but low probability of going into the first triplet state by flipping the spin; the latter state is very vulnerable to destruction by molecular oxygen (for which the ground state is a triplet state). It is also known that β-carotene is an efficient quencher of Chl triplets. Martínez-Junza *et al.* [2008] and Braslavsky and Holzwarth [2012] suggest that the previously unknown function of the D2 branch (see Fig. 5.9b) of the PSII reaction center is to take care of these Chl triplets that are usually formed at high light intensities. A reason for this suggestion is that the β-carotene on this branch is closer to the accessory Chl (Chl_{D1}) than the corresponding one on the D1 branch, and probably it is also positioned in a more suitable direction for energy transfer (Fig. 5.9c).

5.7   Regulation of RuBisCO

Active RuBisCO is a large molecule consisting of four large (L) and four small subunits (S) (see Fig. 5.10 and Section 4.1). The small subunits are coded by chloroplast (**C**) genes and the large subunits by nuclear (**N**) genes. The assembly of the complete enzyme, involving the transport of the large subunits across the nuclear and the plastid membranes, and their proper folding is aided by at least one chaperone; we know that it is not a simple process, but we refrain from describing its details here.

Catalytic activity of RuBisCO requires reversible condensation of CO_2 with an internal lysine (lysyl) residue, which is outside the catalytic site; this leads to the formation of a carbamate (see below), which is stabilized by the catalytically essential Mg^{2+} (Fig. 5.10). Activation of this enzyme requires the nucleus-encoded enzyme RuBisCO activase (RA) for it to work at physiological concentrations of CO_2 [Bowes *et al.*, 1971]. This enzyme is active only when both CO_2 and Mg^{2+} are bound to it.

The activity of RuBisCO is light dependent, and part of the explanation is due to the light-dependency of pH and Mg^{2+} concentration in the stroma. How does it work? Several steps after light absorption lead to the

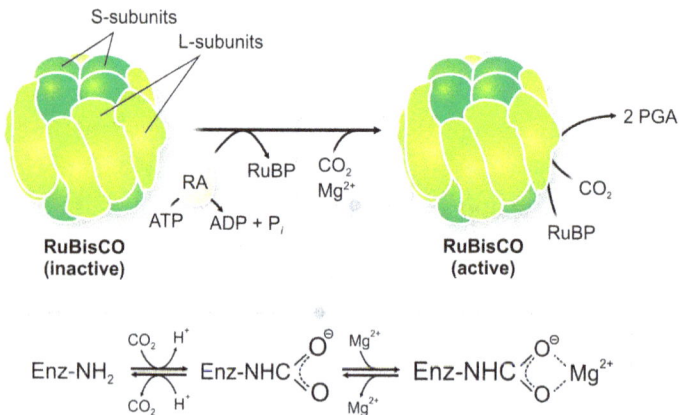

Fig. 5.10.   Schematic representation of RuBisCO activation. The RuBP released during activation was bound to the inactive RuBisCO.

94

pH of the stroma to rise from 7.0 to 8.0 because of proton translocation into the lumen; further, Mg^{2+} ions move out of the thylakoids, increasing their concentration in the stroma. (For role of ions in in regulation, see Kaňa and Govindjee, 2016.) RuBisCO has a high pH optimum, and, thus, becomes "activated" by the addition of CO_2 and Mg^{2+}. However, this is not the whole story. RA, in turn, requires light for its activation, and this is because light increases concentration of ATP, which activates RA, and decreases amount of ADP, which inhibits the RuBisCO.

RuBP has a high affinity for the inactive form of RuBisCO, and thereby keeps the enzyme inactive in the absence of active RA. When RuBisCO is activated by the activase, RuBP bound to the inhibition site is released. For further information, we refer the readers to Portis [1992], Hartman and Harpel [1994], Gutteridge and Jordan [2001], and Hazra *et al.* [2015].

5.8 Redox Regulation of Certain Calvin-Benson Cycle Enzymes by Thioredoxin

In this section, we will focus mainly on plants; for cyanobacteria and algae, see discussion by Tamoi and Shigeoka [2015]. Several of the enzymes of the Calvin-Benson cycle can exist in either of two forms, an active form with reduced sulfur group (i.e., sulfhydryl, —SH), and an inactive form with oxidized disulfide group(s) (—S–S—). The reduction of the inactive form to the active form is carried out by thioredoxins, of which there are five different kinds in *Arabidopsis*. There is some disagreement regarding the function of the various forms, but for the Calvin-Benson cycle the most important forms are thioredoxin *m* (the most abundant form, Okegawa and Motohashi [2015]) and thioredoxin *f*1 [Thormählen *et al.*, 2015]. The disulfide groups of the thioredoxins can be reduced to the sulfhydryl groups either by reduced ferredoxin, catalyzed by ferredoxin-thioredoxin oxidoreductase, or by NADPH *via* mediation of NADPH-dependent thioredoxin reductase C. Many of the enzymes of the Calvin–Benson cycle can in this way be reduced and activated by thioredoxins, but the most important regulation is that of seduheptulose bisphosphatase and fructose bisphosphatase, since the activities of these enzymes often limit the rate of CO_2 assimilation [Kossmann *et al.*, 1994;

Lefebvre *et al.*, 2005; Tamoi *et al.*, 2006]. Uematsu *et al.* [2012] have shown that in tobacco plants, fructose 1,6-bisphosphate aldolase activity may be limiting the assimilation of CO_2. Inactivation of this aldolase takes place when, due to insufficient light, neither reduced ferredoxin nor NADPH is available; this prevents the reactions to run backwards and waste assimilate. Further, this inactivation takes place also under oxidative stress.

5.9 Other Regulatory Mechanisms Related to Photosynthesis

There are several regulation mechanisms of relevance to photosynthesis, other than those discussed above. We will not describe them in any detail here, but we briefly mention below some of the most important ones.

(1) Regulation of stomata implies a delicate balance between the need for letting in CO_2 and restricting the loss of water. For further information, see Schroeder *et al.* [2001]; Murata *et al.* [2015]; and Engineer *et al.* [2016]. CO_2 concentration has long-term effects on the size and frequency of stomata; such measurements on stomata of fossil plants have been used to estimate the past CO_2 content of the atmosphere [Royer, 2001; Wang *et al.*, 2015]. A simple feedback mechanism sets in when photosynthesis proceeds at such high rates that transport out of the leaf or conversion to starch does not keep pace with it. Then the osmotic potential in the green mesophyll cells drops, and they begin to draw water from the guard cells, and then the stomata close [Nikinmaa *et al.*, 2013].

(2) Transcription of a large number of genes related to photosynthesis is regulated in many ways. However, only some of the proteins involved in photosynthesis are manufactured in chloroplasts. Most of the genes that were originally present in cyanobacteria, which had evolved to become chloroplasts, have been exported to the nucleus. Perhaps, this was because the radicals and the molecular oxygen produced in chloroplasts would have made them dangerous locations for DNA. In addition to this, the opposite also took place: some genes were transferred from the nuclei to the plastids [Rodríguez-Moreno *et al.*,

2011]. The reasons why some genes are retained in mitochondria and plastids, and others transferred to the nucleus, have been discussed by Allen [2015] and Johnston and Williams [2016]. Feedback from photosynthesis to genes located in the nucleus (retrograde signaling) is not really known, but is currently a hot topic of research [Pogson *et al.*, 2008; Rüdiger and Oster, 2012; Barajas-López *et al.*, 2013; Gollan *et al.*, 2015; Singh *et al.*, 2015; Crawford *et al.*, 2018].

(3) The roles of ROS, i.e., H_2O_2, $HO_2^{\cdot}$, $HO^{\cdot}$, $^1O_2$ and $O_2^{-\cdot}$ in signaling related to photosynthesis are summarized by Mittler *et al.* [2011]; Schmitt *et al.* [2014]; and Vishwakarma *et al.* [2015].

5.10   Leaf Movement

Leaves of many plants can move in a way that helps in adjusting the energy influx to an optimal value. Many species of *Oxalis* grow in forests where tree canopies intercept most of the light, and their photosynthetic apparatus is adapted to weak light. But sometimes direct sunlight hits these plants through a gap in the canopy. They then fold their leaves very quickly together, a reaction that has been shown to protect the chloroplasts from damage [Björkman and Powles, 1981]. Other plants, such as those of *Lupinus*, are able to change their positions and follow the sun during the day to maximize light exposure [Vogelmann and Björn, 1983].

5.11   Concluding Remarks

As summarized in this chapter, photosynthetic organisms are equipped with a number of regulation systems, which protect them from damage associated with energy overload. They also, to some extent, have the ability to repair damage. For those attempting to utilize the energy of sunlight technologically by using the principles of natural photosynthesis (see Chapter 9), the main challenge may not be in increasing the already efficient basic photochemistry and electron transport, but to make the system robust and durable and functional under variable environmental conditions. In the next chapter we shall outline the global impact of photosynthesis on Earth.

References

Allen, J. F. (2015). Why chloroplasts and mitochondria retain their own genomes and genetic systems: Colocation for redox regulation of gene expression, *Proc. Natl. Acad. Sci. U.S.A.*, 112, pp. 10231–10238.

Allen, J. F. and Mullineaux, C. W. (2004). Probing the mechanism of state transitions in oxygenic photosynthesis by chlorophyll fluorescence spectroscopy, kinetics and imaging. *In* Papageorgiou, G. C. and Govindjee, eds, *Chlorophyll a Fluorescence: A Signature of Photosynthesis* (Springer, Dordrecht), pp. 447–461.

Ananyev, G., Gates, C. and Dismukes, G. C. (2016). The oxygen quantum yield in diverse algae and cyanobacteria is controlled by partitioning of flux between linear and cyclic electron flow within photosystem II, *Biochim. Biophys. Acta*, 1857, pp. 1380–1391.

Ananyev, G., Gates, C., Kaplan, A. and Dismukes, G. C. (2017). Photosystem II-cyclic electron flow powers exceptional photoprotection and record growth in the microalga *Chlorella ohadii*, *Biochim. Biophys. Acta*, 1858, pp. 873–883.

Aro, E.-M. and Andersson, B., eds., (2001). *Regulation of Photosynthesis. Advances in Photosynthesis and Respiration* (Springer, Dordrecht).

Ballottari, M., Alcocer, M. J. P., D'Andrea, C., Viola, D., Ahn, T. K., Petrozza, A., Polli, D., Fleming, G. R., Cerullo, G. and Bassi, R. (2014). Regulation of photosystem I light harvesting by zeaxanthin, *Proc. Natl. Acad. Sci. U.S.A.*, 111, pp. E2431–E2438.

Barajas-López, J. d. D., Blanco, N. E. and Strand, Å. (2013). Plastid-to-nucleus communication, signals controlling the running of the plant cell, *Biochim. Biophys. Acta*, 1833, pp. 425–437.

Bellafiore, S., Barneche, F., Peltier, G. and Rochaix, J.-D. (2005). State transitions and light adaptation require chloroplast thylakoid protein kinase STN7, *Nature*, 433, pp. 892–895.

Björkman, O. and Powles, S. B. (1981). Leaf movement in the shade species Oxalis oregana I. Response to light level and light quality, *Carnegie Inst. Washington Yearb.*, 1980/1981, pp. 59–62.

Bonaventura, C. and Myers, J. E. (1969). Fluorescence and oxygen evolution from *Chlorella pyrenoidosa*, *Biochim. Biophys.Acta*, 189, pp. 366–383.

Bowes, G., Ogren, W. L. and Hageman, R. H. (1971). Phosphoglycolate production catalyzed by ribulose diphosphate carboxylase, *Biochem. Biophys. Res. Commun.*, 45, pp. 716–722.

Braslavsky, S. E. and Holzwarth, A. R. (2012). Role of carotenoids in photosystem II (PSII) reaction centers, *Int. J. Thermophys.*, 33, pp. 2021–2025.

Brinkert, K., De Causmaecker, S., Krieger-Liszkay, A., Fantuzzi, A. and Rutherford, A. W. (2016). Bicarbonate-induced redox tuning in Photosystem II for regulation and protection, *Proc. Natl. Acad. Sci. U.S.A.*, 113, pp. 12144–12149.

Chmeliov, J., Valkunas, L., Krüger, T. P. J., Ilioaia, C. and van Grondelle, R. (2013). Fluorescence blinking of single major light-harvesting complexes, *New J. Phys.*, 15, p. 085007.

Crawford, T., Lehotai, N. and Strand, Å. (2018) The role of retrograde signals during plant stress responses, *J. Exp. Bot.*, 69, pp. 2783–2795.

Crepin, A. and Caffarri, S. (2015). The specific localizations of phosphorylated Lhcb1 and Lhcb2 isoforms reveal the role of Lhcb2 in the formation of the PSI-LHCII supercomplex in *Arabidopsis* during state transitions, *Biochim. Biophys. Acta*, 1847, pp. 1539–1548.

Demmig-Adams, B., Garab, G., Adams, W. and Govindjee, eds (2014). *Non-Photochemical Quenching and Energy Dissipation in Plants, Algae and Cyanobacteria* (Springer, Dordrecht).

Depège, N., Bellafiore, S. and Rochaix, J.-D. (2003). Role of chloroplast protein kinase Stt7 in LHCII phosphorylation and state transition in *Chlamydomonas*, *Science*, 299, pp. 1572–1575.

Derks, A., Schaven, K. and Bruce, D. (2015). Diverse mechanisms for photoprotection in photosynthesis. Dynamic regulation of photosystem II excitation in response to rapid environmental change, *Biochim. Biophys. Acta*, 1847, pp. 468–485.

Duysens, L. N. M. (1972). 3-(3,4-dichlorophenyl)-1,1-dimethylurea (DCMU) inhibition of system II and light-induced regulatory changes in energy transfer efficiency, *Biophys. J.*, 12, pp. 858–863.

Ehleringer, J. R. (2006). Photosynthesis: Physiological and ecological considerations *In* Taiz, L. and Zeiger, E., eds. *Plant Physiology*, 4th ed. (Sinauer Associates, Sunderland), pp. 243–269.

Endo, T. and Asada, K. (2006). Photosystem I and photoprotection: Cyclic electron flow and water-water cycle. *In* Demmig-Adams, B., Adams, W. W. and Mattoo, A. K., eds, *Photoprotection, Photoinhibition, Gene Regulation, and Environment* (Springer Netherlands, Dordrecht), pp. 205–221.

Engineer, C. B., Hashimoto-Sugimoto, M., Negi, J., Israelsson-Nordström, M., Azoulay-Shemer, T., Rappel, W.-J., Iba, K. and Schroeder, J. I. (2016). CO_2 sensing and CO_2 regulation of stomatal conductance: Advances and open questions, *Trends Plant. Sci.*, 21, pp. 16–30.

Esteban, R., Becerril, J. M. and García-Plazaola, J. I. (2009). Lutein epoxide cycle, more than just a forest tale, *Plant Signal. Behav.*, 4, pp. 342–344.

Feyziyev, Y., Deák, Z., Styring, S. and Bernát, G. (2013). Electron transfer from Cyt b_{559} and tyrosine-D to the S_2 and S_3 states of the water oxidizing complex in photosystem II at cryogenic temperatures, *J. Bioenerg. Biomembr.*, 45, pp. 111–120.

Frank, H. A., Bautista, J. A., Josue, J. S. and Young, A. J. (2000). Mechanism of nonphotochemical quenching in green plants: Energies of the lowest excited singlet states of violaxanthin and zeaxanthin, *Biochemistry*, 39, pp. 2831–2837.

Frank, H., Young, A., Britton, G. and Cogdell, R. J., eds., (1999). *The Photochemistry of Carotenoids, Advances in Photosynthesis and Respiration* (Springer, Dordrecht).

Fréchette, E., Wong, C. Y. S., Junker, L. V., Chang, C. Y.-Y. and Ensminger, I. (2015). Zeaxanthin-independent energy quenching and alternative electron sinks cause a decoupling of the relationship between the photochemical reflectance index (PRI) and photosynthesis in an evergreen conifer during spring, *J. Exp. Bot.*, 66, pp. 7309–7323.

Fukuzawa, H., Ogawa, T. and Kaplan, A. (2012). The uptake of CO_2 by cyanobacteria and microalgae, In: Eaton-Rye, J. J., Tripathy, B. C. and Sharkey T. D., eds., *Photosynthesis: Plastid Biology, Energy Conversion and Respiration* (Springer, Dordrecht), pp. 625–650.

Garcia-Plazaola, J. I., Hernandez, A., Errasti, E. and Becerril, J. M. (2002). Occurrence and operation of the lutein epoxide cycle in *Quercus* species, *Funct. Plant. Biol.*, 29, pp. 1075–1080.

Gavel, Y, Steppuhn, J., Herrmann, R. and von Heijne, G. (1991). The "positive-inside rule" applies to thylakoid membrane proteins, *FEBS Lett.*, 282, pp. 41–46.

Goldschmidt-Clermont, M. and Bassi, R. (2015). Sharing light between two photosystems: Mechanism of state transitions, *Curr. Opin. Plant Biol.*, 25, pp. 71–78.

Goss, R. and Lepetit, B. (2015). Biodiversity of NPQ, *J. Plant Physiol.*, 172, pp. 13–32.

Gollan, P. J., Tikkanen, M. and Aro, E.-M. (2015). Photosynthetic light reactions: Integral to chloroplast retrograde signalling, *Curr. Opin. Plant Biol.*, 27, pp. 180–191.

Gutteridge, S. and Jordan, D. B. (2001). Dynamics of photosynthetic CO_2 fixation: Control, regulation and productivity. *In* Aro, E.-M. and Andersson, B., eds, *Regulation of Photosynthesis* (Springer Netherlands, Dordrecht), pp. 297–312.

Hartman, F. C. and Harpel, M. R. (1994). Structure, function, regulation, and assembly of D-ribulose-1,5-bisphosphate carboxylase/oxygenase, *Annu. Rev. Biochem.*, 63, pp. 197–232.

Hazra, S., Henderson, J. N., Liles, K., Hilton, M. T. and Wachter, R. M. (2015). Regulation of ribulose-1,5-bisphosphate carboxylase/oxygenase (Rubisco) activase: Product inhibition, cooperativity, and magnesium activation, *J. Biol. Chem.*, 290, pp. 24222–24236.

Hou, C. X., Pursiheimo, S., Rintamäki, E. and Aro, E. M. (2002). Environmental and metaboliccontrol of LHCII protein phosphorylation: Revealing the mechanismsfor dual regulation of the LHCII kinase, *Plant Cell Environ.*, 25, pp. 1515–1525.

Jahns, P. and Holzwarth, A. R. (2012). The role of the xanthophyll cycle and of lutein in photoprotection of photosystem II, *Biochim. Biophys. Acta*, 1817, pp. 182–193.

Johnston, I. G. and Williams, B. P. (2016). Evolutionary inference across eukaryotes identifies specific pressures favoring mitochondrial gene retention, *Cell Systems*, 2, pp. 101–111.

Kaňa, R. and Govindjee (2016). Role of ions in the regulation of light harvesting, *Front. Plant Sci.*, 7, p. 1849.

Kaňa, R., Kotabová, E., Komárek, O., Šedivá, B., Papageorgiou, G. C., Govindjee and Prášil, O. (2012). The slow S to M fluorescence rise in cyanobacteria is due to a state 2 to state 1 transition, *Biochim. Biophys. Acta*, 1817, pp. 1237–1247.

Kodru, S., Malavath, T., Devadasu, E. , Nellaepalli, S., Stirbet, A., Subramanyam, R. and Govindjee. (2015). The slow S to M rise of chlorophyll a fluorescence induction reflects transition from state 2 to state 1 in the green alga *Chlamydomonas reinhardtii. Photosynth. Res.*, 125, pp. 219–231.

Koroidov, S., Shevela, D., Shutova, T., Samuelsson, G. and Messinger, J. (2014). Mobile hydrogen carbonate acts as proton acceptor in photosynthetic water oxidation, *Proc. Natl. Acad. Sci. U.S.A.*, 111, pp. 6299–6304.

Kossmann, J., Sonnewald, U. and Willmitzer, L. (1994). Reduction of the chloroplastic fructose-1,6-bisphosphatase in transgenic potato plants impairs photosynthesis and plant growth, *Plant J.*, 6, pp. 637–650.

Krüger, T. P. J., Ilioaia, C., Valkunas, L. and van Grondelle, R. (2011a). Fluorescence intermittency from the main plant light-harvesting complex: Sensitivity to the local environment, *J. Phys. Chem. B*, 115, pp. 5083–5095.

Krüger, T. P. J., Ilioaia, C. and van Grondelle, R. (2011b). Fluorescence intermittency from the main plant light-harvesting complex: Resolving shifts between intensity levels, *J. Phys. Chem. B*, 115, pp. 5071–5082.

Kurisu, G., Zhang, H., Smith, J. L. and Cramer, W. A. (2003). Structure of the cytochrome $b_6 f$ complex of oxygenic photosynthesis: Tuning the cavity, *Science*, 302, pp. 1009–1014.

Laisk, A., Oja, V., Eichelmann, H. and Dall'Osto, L. (2014). Action spectra of photosystems II and I and quantum yield of photosynthesis in leaves in State 1, *Biochim. Biophys. Acta*, 1837, pp. 315–325.

Latowski, D., Kuczyńska, P. and Strzałka, K. (2011). Xanthophyll cycle – A mechanism protecting plants against oxidative stress, *Redox Rep.*, 16, pp. 78–90.

Lefebvre, S., Lawson, T., Fryer, M., Zakhleniuk, O. V., Lloyd, J. C. and Raines, C. A. (2005). Increased sedoheptulose-1,7-bisphosphatase activity in transgenic tobacco plants stimulates photosynthesis and growth from an early stage in development, *Plant Physiol.*, 138, pp. 451–460.

Liu, Z., Yan, H., Wang, K., Kuang, T., Zhang, J., Gui, L., An, X. and Chang, W. (2004). Crystal structure of spinach major light-harvesting complex at 2.72 Å resolution, *Nature*, 428, pp. 287–292.

Longoni, P., Douchi, D., Cariti, F., Fucile, G. and Goldschmidt-Clermont, M. (2015). Phosphorylation of the light-harvesting complex II isoform Lhcb2 is central to state transitions, *Plant Physiol.*, 169, p. 2874.

Martínez-Junza, V., Szczepaniak, M., Braslavsky, S. E., Sander, J., Nowaczyk, M., Rogner, M. and Holzwarth, A. R. (2008). A photoprotection mechanism involving the D2 branch in photosystem II cores with closed reaction centers, *Photochem. Photobiol. Sci.*, 7, pp. 1337–1343.

Matsubara, S., Krause, G. H., Aranda, J., Virgo, A., Beisel, K. G., Jahns, P. and Winter, K. (2009). Sun-shade patterns of leaf carotenoid composition in 86 species of neotropical forest plants, *Funct. Plant Biol.*, 36, pp. 20–36.

Matsubara, S., Chen, Y.-C., Caliandro, R., Govindjee and Clegg, R. M. (2011). Photosystem II fluorescence lifetime imaging in avocado leaves: Contributions of the lutein-epoxide and violaxanthin cycles to fluorescence quenching, *J. Photochem. Photobiol. B Biol.*, 104, pp. 271–284.

Minagawa, J. (2011). State transitions—The molecular remodeling of photosynthetic supercomplexes that controls energy flow in the chloroplast, *Biochim. Biophys. Acta*, 1807, pp. 897–905.

Mittler, R., Vanderauwera, S., Suzuki, N., Miller, G., Tognetti, V. B., Vandepoele, K., Gollery, M., Shulaev, V. and Van Breusegem, F. (2011). ROS signaling: The new wave? *Trends Plant. Sci.*, 16, pp. 300–309.

Murata, N. (1969). Control of excitation transfer in photosynthesis I. Light-induced change of chlorophyll a fluoresence in *Porphyridium cruentum*, *Biochim. Biophys. Acta*, 172, pp. 242–251.

Murata, N. (2009). The discovery of state transitions in photosynthesis 40 years ago, *Photosynth. Res.*, 99, pp. 155–160.

Murata, Y., Mori, I. C. and Munemasa, S. (2015). Diverse stomatal signaling and the signal integration mechanism, *Annu. Rev. Plant Biol.*, 66, pp. 369–392.

Nikinmaa, E., Hölttä, T., Hari, P., Kolari, P., Mäkelä, A., Sevanto, S. and Vesala, T. (2013). Assimilate transport in phloem sets conditions for leaf gas exchange, *Plant Cell Environ.*, 36, pp. 655–669.

Okegawa, Y. and Motohashi, K. (2015). Chloroplastic thioredoxin m functions as a major regulator of Calvin cycle enzymes during photosynthesis *in vivo*, *Plant J.*, 84, pp. 900–913.

Papageorgiou, G. and Govindjee (2011). Photosystem II fluorescence: Slow changes - Scaling from the past, *J. Photochem. Photobiol. B Biol.*, 104, pp. 258–270.

Papageorgiou, G. and Govindjee (2014). The non-photochemical quenching of the electronically excited state of chlorophyll a in plants: Definitions, timelines, viewpoints, open questions. In: Demmig-Adams, B., Garab, G., Adams, W. and Govindjee, eds, *Non-photochemical Quenching and Energy Dissipation in Plants, Algae and Cyanobacteria* (Springer, Dordrecht), pp. 1–44.

Pogson, B. J., Woo, N. S., Förster, B. and Small, I. D. (2008). Plastid signalling to the nucleus and beyond, *Trends Plant. Sci.*, 13, pp. 602–609.

Polívka, T., Herek, J. L., Zigmantas, D., Åkerlund, H.-E. and Sundström, V. (1999). Direct observation of the (forbidden) S_1 state in carotenoids, *Proc. Natl. Acad. Sci. U.S.A.*, 96, pp. 4914–4917.

Polívka, T. and Sundström, V. (2004). Ultrafast dynamics of carotenoid excited states–From solution to natural and artificial systems, *Chem. Rev.*, 104, pp. 2021–2072.

Portis, A. R. (1992). Regulation of bibulose 1,5-bisphosphate carboxylase/oxygenase activity, *Annu. Rev. Plant Physiol. Plant Mol. Biol.*, 43, pp. 415–437.

Pribil, M., Pesaresi, P., Hertle, A., Barbato, R. and Leister, D. (2010). Role of plastid protein phosphatase TAP38 in LHCII dephosphorylation and thylakoid electron flow, *PLOS Biol.*, 8, p. e1000288.

Powles, S. B. and Bjorkman, O. (1981). Leaf movement in the shade species *Oxalis oregana* II. The role in protection against injury by intense light., *Carnegie Inst. Washington Yearb.*, 1980/1981, pp. 63–66.

Pursiheimo, S., Martinsuo, P., Rintamäki, E. and Aro, E.-M. (2003). Photosystem II protein phosphorylation follows four distinctly different regulatory patterns induced by environmental cues, *Plant Cell Environ.*, 26, pp. 1995–2003.

Rodríguez-Moreno, L., González, V. M., Benjak, A., Martí, M. C., Puigdomènech, P., Aranda, M. A. and Garcia-Mas, J. (2011). Determination of the melon chloroplast and mitochondrial genome sequences reveals that the largest reported mitochondrial genome in plants contains a significant amount of DNA having a nuclear origin, *BMC Genom.*, 12, pp. 424–424.

Royer, D. L. (2001). Stomatal density and stomatal index as indicators of paleoatmospheric CO_2 concentration, *Rev. Palaeobot. Palyn.*, 114, pp. 1–28.

Rüdiger, W. and Oster, U. (2012). Intracellular signaling from plastids to the nucleus. *In* Eaton-Rye, J. J., Tripathy, B. C. and Sharkey, T. D., eds, *Photosynthesis: Plastid Biology, Energy Conversion and Carbon Assimilation* (Springer Netherlands, Dordrecht), pp. 175–190.

Schlau-Cohen, G. S., Yang, H.-Y., Krüger, T. P. J., Xu, P., Gwizdala, M., van Grondelle, R., Croce, R. and Moerner, W. E. (2015). Single-molecule identification of quenched and unquenched states of LHCII, *J. Phys. Chem. Lett.*, 6, pp. 860–867.

Schmidt, L., Christensen, H. E. M. and Harris, P. (2006). Structure of plastocyanin from the cyanobacterium *Anabaena variabillis*, *Acta Crystallogr.*, 62, pp. 1022–1029.

Schmitt, F.-J., Renger, G., Friedrich, T., Kreslavski, V. D., Zharmukhamedov, S. K., Los, D. A., Kuznetsov, V. V. and Allakhverdiev, S. I. (2014). Reactive oxygen species: Re-evaluation of generation, monitoring and role in stress-signaling in phototrophic organisms, *Biochim. Biophys. Acta*, 1837, pp. 835–848.

Schroeder, J. I., Allen, G. J., Hugouvieux, V., Kwak, J. M. and Waner, D. (2001). Guard cell signal transduction, *Annu. Rev. Plant Physiol. Plant Mol. Biol.*, 52, pp. 627–658.

Schörner, M., Beyer, S. R., Southall, J., Cogdell, R. J. and Köhler, J. (2015). Multi-level, multi time-scale fluorescence intermittency of photosynthetic LH2 complexes: A precursor of non-photochemical quenching? *J. Phys. Chem. B*, 119, pp. 13958–13963.

Shapiguzov, A., Ingelsson, B., Samol, I., Andres, C., Kessler, F., Rochaix, J.-D., Vener, A. V. and Goldschmidt-Clermont, M. (2010). The PPH1 phosphatase is specifically involved in LHCII dephosphorylation and state transitions in *Arabidopsis*, *Proc. Natl. Acad. Sci. U.S.A.*, 107, pp. 4782–4787.

Shevela, D., Noring, B., Koroidov, S., Shutova, T., Samuelsson, G. and Messinger, J. (2013). Efficiency of photosynthetic water oxidation at ambient and depleted levels of inorganic carbon, *Photosynth. Res.*, 117, pp. 401–412.

Shinopoulos, K. E. and Brudvig, G. W. (2012). Cytochrome b_{559} and cyclic electron transfer within photosystem II, *Biochim. Biophys. Acta*, 1817, pp. 66–75.

Shutova, T., Kenneweg, H., Buchta, J., Nikitina, J., Terentyev, V., Chernyshov, S., Andersson, B., Allakhverdiev, S. I., Klimov, V. V., Dau, H., Junge, W. and Samuelsson, G. (2008). The photosystem II-associated Cah3 in *Chlamydomonas* enhances the O_2 evolution rate by proton removal, *EMBO J.*, 27, pp. 782–791.

Singh, R., Singh, S., Parihar, P., Singh, V. P. and Prasad, S. M. (2015). Retrograde signaling between plastid and nucleus: A review, *J. Plant Physiol.*, 181, pp. 55–66.

Su, X., Ma, J., Wei, X., Cao, P., Zhu, D., Chang, W., Liu, Z., Zhang, X. and Li, M. (2017). Structure and assembly mechanism of plant $C_2S_2M_2$-type PSII-LHCII supercomplex, *Science*, 357, pp. 815–820.

Tamoi, M., Nagaoka, M., Miyagawa, Y. and Shigeoka, S. (2006). Contribution of fructose-1,6-bisphosphatase and sedoheptulose-1,7-bisphosphatase to the photosynthetic rate and carbon flow in the Calvin cycle in transgenic plants, *Plant Cell Physiol.*, 47, pp. 380–390.

Tamoi, M. and Shigeoka, S. (2015). Diversity of regulatory mechanisms of photosynthetic carbon metabolism in plants and algae, *Biosci. Biotechnol. Biochem.*, 79, pp. 870–876.

Thormählen, I., Meitzel, T., Groysman, J., Öchsner, A. B., von Roepenack-Lahaye, E., Naranjo, B., Cejudo, F. J. and Geigenberger, P. (2015). Thioredoxin $f1$ and NADPH-dependent thioredoxin reductase C have overlapping functions in regulating photosynthetic metabolism and plant growth in response to varying light conditions, *Plant Physiol.*, 169, pp. 1766–1786.

Tikhonov, A. N. (2014). The cytochrome b_6f complex at the crossroad of photosynthetic electron transport pathways, *Plant Physiol. Biochem.*, 81, pp. 163–183.

Uematsu, K., Suzuki, N., Iwamae, T., Inui, M. and Yukawa, H. (2012). Increased fructose 1,6-bisphosphate aldolase in plastids enhances growth and photosynthesis of tobacco plants, *J. Exp. Bot.*, 63, pp. 3001–3009.

Umena, Y., Kawakami, K., Shen, J.-R. and Kamiya, N. (2011). Crystal structure of oxygen-evolving photosystem II at a resolution of 1.9 Å, *Nature*, 473, pp. 55–60.

Vogelmann, T. C. and Björn, L. O. (1983). Response to directional light by leaves of a sun-tracking lupine (*Lupinus succulentus*), *Physiol. Plantarum*, 59, pp. 533–538.

Wang, Y., Momohara, A., Wang, L., Lebreton-Anberrée, J. and Zhou, Z. (2015). Evolutionary history of atmospheric CO_2 during the late Cenozoic from fossilized *Metasequoia* needles, *PLoS ONE*, 10, p. e0130941.

Wydrzynski, T. and Satoh, K., eds., (2005). *Photosystem II: The Light-Driven Water: Plastoquinone Oxidoreductase* (Springer, Dordrecht).

Vishwakarma, A., Tetali, S. D., Selinski, J., Scheibe, R. and Padmesree, K. (2015). Importance of the alternative oxidase (AOX) pathway in regulating cellular redox and ROS homeostasis to optimize photosynthesis during restriction of the cytochrome oxidase pathway in *Arabidopsis thaliana*, *Ann. Bot.*, 116, pp. 555–569.

Zito, F., Finazzi, G., Delosme, R., Nitschke, W., Picot, D. and Wollman, F. A. (1999). The Qo site of cytochrome $b_6 f$ complexes controls the activation of the LHCII kinase, *EMBO J.*, 18, pp. 2961–2969.

Chapter 6

Photosynthesis and Our Planet

6.1 Oxygenation of the Earth's Atmosphere

There is no consensus regarding when oxygenic photosynthesis started on our planet. Most agree that there was a large and rather abrupt increase in atmospheric oxygen around 2.4–2.3 Ga (billion years) ago, during the *Great Oxidation Event* (GOE) (see [Blaustein, 2016; Luo *et al.*, 2016; Gumsley *et al.*, 2017] and Fig. 6.1a). However, oxygenic photosynthesis must have been around much earlier than this time [Planavsky *et al.*, 2014], since the first oxygen produced has been suggested to be consumed by the oxidation of the accumulated reduced iron and sulfur compounds, methane, and hydrogen. By comparing the sequences of a large number of cyanobacterial genes, Schirrmeister *et al.* [2015] have estimated that the evolutionary divergence between *Gloeobacter* (a genus of cyanobacteria, lacking thylakoids) and all other cyanobacteria must have taken place some time between 3.85 and 2.92 Ga ago.

Until recently, it seemed obvious to most that no life could have survived when the Earth was being bombarded with meteorites, peaking around 3.9 Ga ago and ending around 3.72 Ga ago [Fernandes *et al.*, 2013]. However, a current view [Bottke *et al.*, 2017; Zellner, 2017] is that this bombardment was drawn out in time, and that there were always some places on the planet where organisms could have survived [Abramov and Mojzsis, 2009]. Moreover, Tashiro *et al.* [2017] have interpreted the graphite found in 3.95 Ga old sedimentary rocks to be of biological origin. From geochemical indicators, Crowe *et al.* [2013] have deduced that free oxygen was present ~3 Ga ago, and

Photosynthesis: Solar Energy for Life by Dmitry Shevela, Lars Olof Björn and Govindjee
© 2018, published by World Scientific Publishing Co. Pte. Ltd. ISBN: 978-981-3223-10-3.

Riding *et al.* [2014] have noted that at least 10.25 µM O_2 was present in seawater ~2.8 Ga ago. Further, Satkoski *et al.* [2015] have concluded that oxygenic photosynthesis existed as far back as 3.2 Ga ago. On the other hand, another study had suggested the presence of oxygenic photosynthesis even earlier, i.e., 3.7 Ga ago [Rosing and Frei, 2004], but this estimate remains questionable [Lyons *et al.*, 2014; Fischer *et al.*, 2016]. However, the conclusion of Satkoski *et al.* [2015] implies that this very complicated process had evolved within 250 million years from the start of the first life! Finally, Dodd *et al.* [2017] have suggested that putative fossilized microorganisms, found in ferruginous sedimentary rocks, and identified as seafloor-hydrothermal vent-related precipitates, are possibly 4.28 Ga old! If this holds true, it would give about 0.5 Ga for the first living organisms to evolve into the very first oxygenic organisms. In contrast to the above idea, Ward *et al.* [2016] have argued that oxygenic photosynthesis evolved shortly before the GOE, i.e., ~2.4 Ga ago.

The GOE may have quickly, but temporarily, led to an oxygen content close to the present one [Bekker and Holland, 2012; Harada *et al.*, 2015], but again "quickly" (i.e., over *ca.* 100,000 years) the level dropped to *ca.* 3% of the present atmospheric level (PAL). Many organisms, including all the animals (e.g., Metazoa), need oxygen, but opinions differ as to the precise time the oxygen content had reached the level necessary to sustain the first animal life. Three percent of the present oxygen level, i.e., about 8 µM, could have been sufficient for the first animals, as it is for a modern sponge [Mills *et al.*, 2014]. Molecular clock studies date the origin of Metazoa to 750–800 million years (Ma) ago, roughly coinciding with the evidence from geochemical proxies that oxygen levels, during this time, rose from less than 0.1% PAL to perhaps 1–3% PAL O_2 [Erwin, 2015].

6.2 Protection: Ozone Was Formed From Oxygen and It Protects Us From the UV Radiation

The young Sun radiated shortwave UV-C more intensely than the Sun does today [Ribas *et al.*, 2005, 2010; Claire *et al.*, 2012], but the most energetic photons would hardly have reached the surface of the Earth. At that time, the Sun emitted UV-B and long-wave UV-C radiation at a slightly lower rate than it does today [Claire *et al.*, 2012], but on the other hand, they were only very slightly weakened before reaching the ground or the ocean surface [Thomassot *et al.*, 2015]. During the Archaean era (4.0 to 2.5 Ga ago) there was essentially no ozone present that could block UV-B radiation, but according to Domagal-Goldman [2008] an organic haze provided a certain protection to the biosphere during the mid-Archaean era. Most of the life, at that time, was still restricted to the aquatic environment, where additional protection was, most likely, provided by dissolved substances or precipitates, such as grains of ferrihydrite [Gauger *et al.*, 2015].

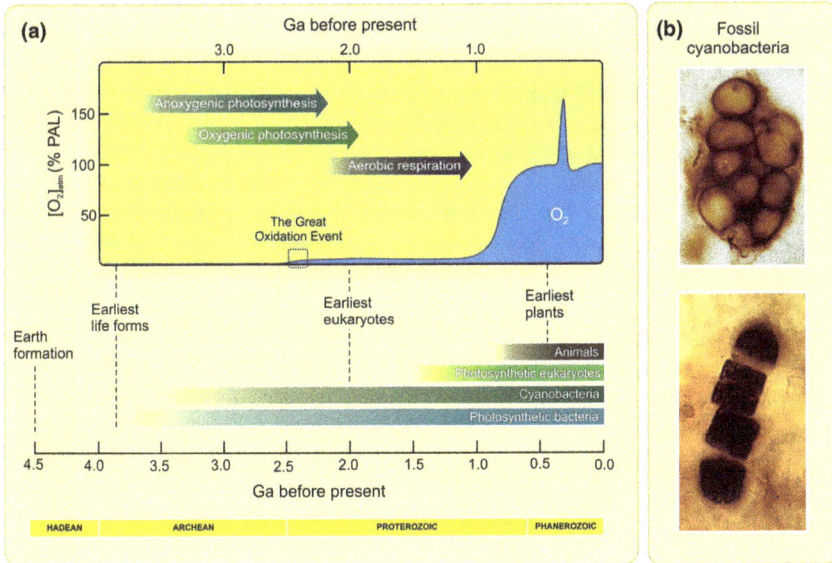

Fig. 6.1. The role of oxygenic photosynthesis in changing the concentration of atmospheric O_2, the evolution of life and evolution of some metabolic pathways during Earth's history. **(a)** Schematic view of the relationship between critical events in the evolution of life and oxygenic photosynthesis, as well as with the concentration of atmospheric O_2 (in % of PAL) through geological times in billions of years (Ga). These relationships, the times, and concentration curve of atmospheric O_2 are approximations, based on a large body of data (see, e.g., [Kump, 2008; Falkowski, 2011; Farquhar *et al.*, 2011; Hohmann-Marriott and Blankenship, 2011; Schopf, 2011; Satkoski *et al.*, 2015; Blaustein, 2016; Fischer *et al.*, 2016; Gumsley *et al.*, 2017]). Here we show only some selected events of evolutionary diversification and the origin of some organisms. **(b)** Two representative fossil cyanobacteria from the ~0.85-Ga-old Bitter Springs Chert (central Australia). *Top*: a non-mobile colonial chroococcacean cyanobacterium (of coccoida form). *Bottom*: a filamentous cyanobacterium *Palaeolyngbya* (Oscillatoriaceae). Interestingly, the oscillatoriaceaen cyanobacteria have changed very little over the last thousands of millions of years [Schopf, 2011]. The photographs of fossil cyanobacteria were provided by J. William Schopf. Modified from Shevela *et al.* [2013].

With the GOE a very effective screen for radiation, deleterious to DNA, must have quickly developed. This was the stratospheric ozone (O_3) shield, formed photochemically from oxygen, after absorption of UV-C radiation. By coincidence, the absorption spectrum of ozone has a peak at the same wavelength as that of DNA. Even the low amount of oxygen, say 1% of the present that first accumulated in the atmosphere, resulted in much more than 1% of the present ozone, although the ozone layer had its maximum at a lower elevation than the present one [Segura *et al.*, 2003].

6.3 Earth Temperature Over Time and the Effect of the Biosphere on It

6.3.1 *Influence of the Sun*

It is remarkable how little the Sun's radiation has varied over billions of years, although it has not been completely constant. Around the time life originated, the emission, by the Sun, in the short-wavelength UV part of the spectrum was much higher [Ribas *et al.*, 2005; Claire *et al.*, 2012], while the total output was 30% lower than it is today [Gough, 1981; Tajika, 2003]. This information has been deduced from the solar models and through comparison with other similar stars at an earlier stage of development. The Stefan-Boltzmann Law [LoPresto and Hagoort, 2011] states that the heat radiation from a body is proportional to the absolute temperature (T) raised to the fourth power, i.e., T^4. The heat reaching the surface of the Earth from the planet's interior is now much less than the heat reaching it from the Sun. How easily the incoming heat is radiated back to the space depends on the surface properties and on the atmosphere; this has been changing over time, but, for simplicity, we shall neglect these changes for now. Then an output from the Sun, lower by 30% than it is today, would have led to an equilibrium only if the outgoing radiation from the Earth was also 30% lower, i.e., 70% of the present Earth radiation. In other words, if the temperature with 100% sunlight is 286.5 K (13.3°C), then with 30% less sunlight, it would be $286.5 \times 0.7^{0.25}$ K = 262.1 K, i.e., 24.4 K lower, far below freezing. Going back only to the Cambrian period (541 to 485.4 Ma ago) and with the Sun's energy output lower by 4.5% than now leads us to estimate that the temperature was 3.3°C lower than it is now.

The distance between the Earth and the Sun is increasing over time due to several factors discussed by Zhang *et al.* [2010]. Unfortunately, various ways of estimating the rate of this change do not lead to the same value. One of the higher values for the Phanerozoic that is still realistic is 15 m per century [Krasinsky and Brumberg, 2004; Zhang *et al.*, 2010]. This corresponds to an Earth–Sun distance, 3 Ga ago, of 99.7% of the present. In our opinion, we can neglect the change in distance to the Sun when considering its effect on the temperature of our Earth, even if the Sun–Earth distance would have increased somewhat more than 15 m per century during the Archaean era [Zhang *et al.*, 2010].

6.3.2 *Influence of the Earth's atmosphere*

The changes in the atmosphere, and the changes in the greenhouse effect, associated with them, are much more important for the temperature than the change in Earth-Sun distance. In the distant past, the higher heat flux from the Earth's interior must have also played a role in keeping the Earth warm, The heat from the Earth's interior, which is now much smaller than the influx of solar energy, is due mainly to two factors: Decay of radioactive elements synthesized before the formation of the Earth [Fatuzzo and Adams, 2015], and gravitational energy released as heat when iron and other heavy elements sank toward the center of the planet. Both these heating effects have diminished over time.

"Greenhouse gases" absorb the heat radiation from the Earth surface and reradiate some of it to the surface instead of letting it escape to space. Methane is now a minor (although not unimportant) greenhouse gas component of the Earth's atmosphere, but in the past there was more of it, partly generated by Archaea [Nisbet and Fowler, 2011], the "third domain" in the living world discovered by Carl Woese [Woese *et al.*, 1990]. Methane is a much more efficient greenhouse gas than CO_2 because of the positions of its absorption bands. If methane had not been removed by its oxidation by oxygen, produced by photosynthesis, and if CO_2 had not been removed by photosynthetic assimilation at the appropriate rate, the present Earth would have been too hot for survival, because both methane and CO_2 are greenhouse gases. Increased amount of these gases in the atmosphere would probably have resulted in a

111

"runaway greenhouse effect" (due to positive temperature feedback), which, we believe, took place on the lifeless Venus. Thus, water would have evaporated from the surface of the Earth. This would have happened despite the fact that heat is consumed in evaporation. Of a greater importance here is the increased water vapor which is a greenhouse gas, too. And as temperature would have increased, CO_2 would have been released from carbonate in rocks and water, adding even more greenhouse gas to the atmosphere. The danger of this happening is not completely gone, especially with the current high use of fossil fuels and the concomitant increase in atmospheric CO_2, and the risk of release of methane from methane clathrate (an association between methane and water in the ocean) [Mestdagh *et al.*, 2017; Ruppel and Kessler, 2017], as well as from the methane stored in boreal permafrost [Helbig *et al.*, 2017].

CO_2 cannot be removed from the atmosphere by photosynthesis alone. If organic carbon accumulates in the biosphere, most of it will soon get converted to CO_2, by respiration (including microbial respiration when organisms decay). Therefore, processes, such as the formation of coal and oil, which remove organic material from the biosphere are, in a long-term perspective, important for removing CO_2 from the atmosphere. Plants have also contributed to the removal of organic material from the biosphere in other ways than by photosynthesis. When plants appeared on land, weathering and erosion of its surface increased [Berner and Kothavala, 2001]. This had the double effect of enclosing organic material in inorganic sediment, and in providing increased nutrients for cyanobacteria and algae in the ocean as well as in fresh water bodies. In addition, increased weathering caused by plants must have led to an increase in the conversion of CO_2 and silicate to carbonate.

Plate tectonics is another process that affects the CO_2 content of the atmosphere. When oceanic plates of the Earth's crust are pushed under the continents, carbon "disappears" for a long time, until it eventually emerges again in volcanic areas, not always during violent volcanic eruptions, but more gradually. *What is swept under the carpet eventually turns up again!*

During the Cambrian period (541 to 485 Ma ago) the atmospheric concentration of CO_2 was much higher than it is now [Berner, 2006], but this cannot explain why the Earth was so warm despite a fainter Sun than the

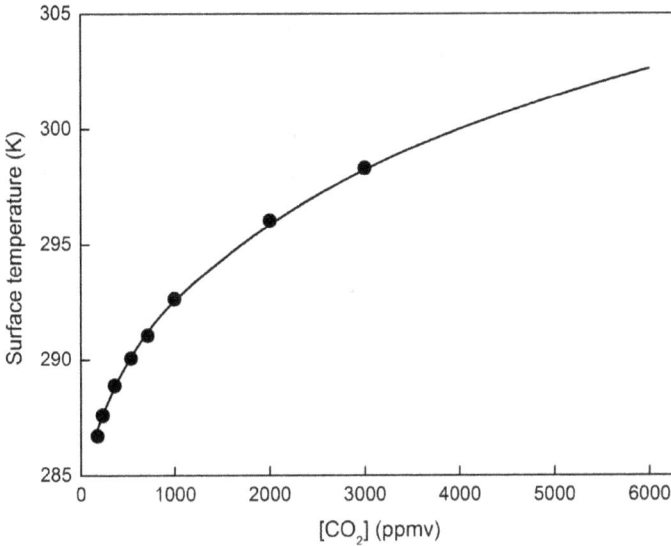

Fig. 6.2. The relationship between global average temperature of the Earth surface and CO_2 concentration in the atmosphere according to Kothavala *et al.* [1999]. The black circles correspond to values calculated using the third version of the "general circulation model," CCM3. The solid line follows the formula $T = 0.647(\ln[CO_2])^2 - 4.5566\ln[CO_2] + 293.37$ K, where T is absolute temperature and $[CO_2]$ the CO_2 concentrations in ppmv (parts per million on a volume basis).

present one. The reason is that the absorption spectrum of dilute gases in the atmosphere is confined to relatively narrow spectral bands, and when absorption within these bands has reached almost 100%, further increase in the concentration of the same gas does not have much effect. Kothavala *et al.* [1999] have provided a quantitative relationship between CO_2 concentration and global surface temperature (Fig. 6.2) (when other factors were assumed to be constant).

Although extrapolation above 3,000 ppm CO_2 is not strictly valid, it provides a rough estimate of the effect of 5,000 ppm CO_2. The effect of an increase of CO_2 from 280 to 5,000 ppm, thus obtained, is equivalent to an increase in temperature by 14.7°C. Strictly speaking it is the total "column" up to the top of the atmosphere that is important, and we should really use the partial pressure of CO_2. We can roughly set 1 ppmv to correspond to 1.5 millionth of a bar, taking into account that the molecular weight of CO_2 (44) is higher than the average (29) for all the atmospheric gases.

113

If we go further back in time to the Proterozoic (2.5 to 0.541 Ga ago) and Archaean (4.0 to 2.5 Ga ago) eons, the lower emission from the Sun could not have been compensated by even much higher content of CO_2. As we saw in the previous section, the fainter young Sun would result in a temperature, which is 24.4 K lower than it is now. Greenhouse gases, other than CO_2, have been suggested to fill in the parts of the Earth's emission spectrum where CO_2 does not absorb. One possibility is N_2O, which is an intermediate in denitrification from NO_3^- to N_2. The final step from N_2O to N_2 requires components of nitrite reductases (copper or iron), which, before the GOE, were present in a form that could have been taken up by the organisms. After the GOE, sulfide minerals, on the continents, underwent oxidative weathering and the sulfates were washed into the ocean. Deep down inside the Earth, conditions were still reducing, and the sulfate was reduced to hydrogen sulfide, which must have precipitated copper and iron to unavailable sulfides. Thus, denitrification would have been halted at N_2O, which would have accumulated in the atmosphere [Roberson *et al.*, 2011].

Another possibility for explaining the high temperature of the Earth in the past, while the Sun was fainter, is the presence of methane [Pavlov *et al.*, 2003; Lazar *et al.*, 2012]; it is formed by not only the methanogens (Archaea), but also abiogenically. Even today, methane continues to rise from the deep sea vents, located at the oceanic ridges, where the oceanic plates are formed [Konn *et al.*, 2015]. Biogenic methane production is thought to have taken place at a much higher rate, before the GOE, because of the better availability of nickel [Konhauser *et al.*, 2009; Konhauser *et al.*, 2015], an element necessary for this process. Nickel is part of the enzyme methyl-coenzyme M reductase, which catalyzes the rate-limiting step of methane formation [Wongnate *et al.*, 2016].

An accurate modeling of heating of the Earth in the deep past is difficult, and one must also take into account the heating from within the Earth, by radioactivity [Dye, 2015], and the sinking of denser constituents toward the center [Geng and Wang, 2015]. During the sinking of denser components, their gravitational energy is converted to heat. Conversion of rotational energy into heat by tidal deceleration of the Earth's rotation seems to have played a smaller role in keeping the Earth warm [Lambeck,

1975; Furlong and Chapman, 2013], although we expect this value to have been higher in the past, when the Moon was closer, because a smaller distance would mean greater gravitational attraction and stronger tidal forces. The heat generated from radioactivity of $^{238}U + {}^{232}Th + {}^{40}K$ was also higher in the past before much of the radioactivity had decayed. The current total net radiation by the Earth to space is 44–47 TW (4.4×10^{13}–4.7×10^{13} W) [Pollack *et al.*, 1993; Furlong and Chapman, 2013; Schwarzschild, 2014], where 14.3 TW (1.43×10^{13} W) of this appears to be heat from the interior remaining from the past heating processes. For comparison, the incoming solar radiation amounts to, on the average, 173,000 TW (1.73×10^{17} W).

The spread of plants on land during the Ordovician (488.3–443.7 Ma ago) must have caused increased weathering of silicate rocks and the conversion of atmospheric CO_2 to carbonate. This, in turn, must have caused cooling and glaciation [Lenton *et al.*, 2012; Lenton and Daines, 2017]. Plants have had important effects on our climate ever since [Boyce and Lee, 2017].

6.4 Conclusion

During the period between 3.7 and 3.2 Ga ago, some ancient cyanobacteria-like organisms must have evolved a molecular apparatus capable of utilizing H_2O as a substrate (i.e., as a source of electrons and protons) for the reduction of CO_2, using sunlight to drive this reaction. Since the metabolic waste product of this process was molecular oxygen, the first oxygenic photosynthesizers had the great advantage in "poisoning" their competitors with "toxic" O_2. As a result, most of the ancient anaerobic organisms did not adapt to the presence of O_2 and faced extinction as O_2 accumulated in the environment [Sleep and Bird, 2008]. Some anaerobic bacteria managed to survive in the remaining ecological O_2 free niches. Others, the most successful of the surviving organisms, managed not only to develop protective mechanisms against O_2, but also to evolve highly efficient respiratory processes, which utilized O_2 as the terminal electron acceptor for "biochemical burning" of the organic compounds. The latter allowed the liberation of at least 10–15 times more free energy from the organic matter than through O_2-free processes [Renger, 1983;

Peschek *et al.*, 2011]. For example, 74 kJ mole^{-1} is released in the fermentation of glucose to CO_2 and ethanol, and more than 2800 kJ mole^{-1} in the complete oxidation of glucose in aerobic respiration. Already about 2.3 Ga ago, the amount O_2 in the atmosphere, produced by cyanobacteria, was high enough to form an O_3 layer, which shielded the early Earth's biosphere from the highly damaging UV radiation from the Sun [Olson and Blankenship, 2004; Kump, 2008]. As a consequence, this permitted organisms "to occupy" the terrestrial environment, and thus, significantly increase genomic and metabolic complexity of life [Payne *et al.*, 2011].

Thus, over the past 3 Ga, the successful version of a photosynthetic electron-transport apparatus has survived and enriched our planet with O_2 (for a recent viewpoint on the amount of oxygen production by photosynthesis, see [Borisov and Björn, 2018]). All of the events described above had a dramatic effect on the evolution of life, as we know it today. How did these first oxygenic organisms that changed the world, and tuned evolution of life into an aerobic direction, look like? We do not know for sure, but it is possible that they looked very similar to the one of the two fossil cyanobacteria shown in Fig. 6.1b (for further details, see [Schopf, 2011]).

Life started on Earth almost as quickly as the planet became habitable. It is very likely that there are many planets in the Universe that have conditions compatible with life. It is reasonable to assume that life in some form evolves (or has evolved) also on such planets. But whether the life evolves (or has evolved) to such sophistication as here on Earth may be much less probable. Perhaps life would not have persisted on Earth if it had not started rather quickly. Without a remarkable interaction between geological and biological processes, as plate tectonics and assimilation of CO_2, the evolution of land plants (see, e.g., Kiang *et al.* [2007] and Kiang [2008] for interesting speculations about plants in other worlds), the weathering caused by them, and burial of organic carbon below sediments, our climate would have been very different. Arguments have been advanced for the view that the probability is very low for the various processes starting at a time and taking place at a rate that leads to habitability for as long a time as our biosphere has enjoyed [Chopra and Lineweaver, 2016].

References

Abramov, O. and Mojzsis, S. J. (2009). Microbial habitability of the Hadean Earth during the late heavy bombardment, *Nature*, 459, pp. 419–422.

Bekker, A. and Holland, H. D. (2012). Oxygen overshoot and recovery during the early Paleoproterozoic, *Earth Planet. Sci. Lett.*, 317, pp. 295–304.

Berner, R. A. (2006) GEOCARBSULF: A combined model for Phanerozoic atmospheric O_2 and CO_2, *Geochim. Cosmochim. Acta*, 70, pp. 5653–5664.

Berner, R. A. and Kothavala, Z. (2001). GEOCARB III: A revised model of atmospheric CO_2 over phanerozoic time, *Am. J. Sci.*, 301, pp. 182–304.

Blaustein, R. (2016). The Great Oxidation Event: Evolving understandings of how oxygenic life on Earth began, *BioScience*, 66, pp. 189–195.

Borisov, A. Y. and Björn, L. O. (2018). On oxygen production by photosynthesis: A viewpoint, *Photosynthetica*, 56, pp. 44–47.

Bottke, W. F., Vokrouhlický, D., Minton, D., Nesvorný, D., Morbidelli, A., Brasser, R., Simonson, B. and Levison, H. F. (2017). An Archaean heavy bombardment from a destabilized extension of the asteroid belt, *Nature*, 485, pp. 78–81.

Boyce, C. K. and Lee, J.-E. (2017). Plant evolution and climate over geological timescales, *Annu. Rev. Earth Planet. Sci.* 45, pp. 61–87.

Chopra, A. and Lineweaver, C. H. (2016). The case for a gaian bottleneck: The biology of habitability, *Astrobiology*, 16, pp. 7–22.

Claire, M. W., Sheets, J., Cohen, M., Ribas, I., Meadows, V. S. and Catling, D. C. (2012). The evolution of solar flux from 0.1 nm to 160 μm: Quantitative estimates for planetary studies, *Astrophys. J.*, 757, p. 95.

Crowe, S. A., Dossing, L. N., Beukes, N. J., Bau, M., Kruger, S. J., Frei, R. and Canfield, D. E. (2013). Atmospheric oxygenation three billion years ago, *Nature*, 501, pp. 535–538.

Dodd, M. S., Papineau, D., Grenne, T., Slack, J. F., Rittner, M., Pirajno, F., O'Neil, J. and Little, C. T. S. (2017). Evidence for early life in Earth's oldest hydrothermal vent precipitates, *Nature*, 543, pp. 60–64.

Domagal-Goldman, S. D., Kasting, J. F., Johnston, D. T. and Farquhar, J. (2008). Organic haze, glaciations and multiple sulfur isotopes in the Mid-Archean Era, *Earth Planet Sci Lett.*, 269, pp. 29–40.

Dye, S. T. (2015). Geo-neutrinos: Recent developments, *Nucl. Part. Phys. Proc.*, 265, pp. 114–116.

Erwin, D. H. (2015). Early metazoan life: Divergence, environment and ecology, *Philos. Trans. R. Soc. Lond.*, B, 370, p. 20150036.

Falkowski, P. (2011). The biological and geological contingencies for the rise of oxygen on Earth, *Photosynth. Res.*, 107, pp. 7–10.

Farquhar, J., Zerkle, A. and Bekker, A. (2011). Geological constraints on the origin of oxygenic photosynthesis, *Photosynth. Res.*, 107, pp. 11–36.

Fatuzzo, M. and Adams, F. C. (2015). Distributions of long-lived radioactive nuclei provide by star-forming environments, *Astrophys. J.*, 813, p. 55.

Fernandes, V. A., Fritz, J., Weiss, B., Garrick-Bethell, I. and Shuster, D. L. (2013). The bombardment history of the Moon as recorded by 40Ar-39Ar chronology, *Meteoritics Planet. Sci.*, pp. 1–29.

Fischer, W. W., Hemp, J. and Johnson, J. E. (2016). Evolution of oxygenic photosynthesis, *Annu. Rev. Earth Planet. Sci.*, 44, pp. 647–683.

Furlong, K. P. and Chapman, D. S. (2013). Heat flow, heat generation, and the thermal state of the lithosphere, *Annu. Rev. Earth Planet. Sci.*, 41, pp. 385–410.

Gauger, T., Konhauser, K. and Kappler, A. (2015). Protection of phototrophic iron(II)-oxidizing bacteria from UV irradiation by biogenic iron(III) minerals: Implications for early Archean banded iron formation, *Geology*, 43, pp. 1067–1070.

Geng, Y. and Wang, J. H. (2015). Research on calculation methods of differentiation energy during the formation and evolution of the earth, *Chin. J. Geophys.*, 58, pp. 3530–3539.

Gough, D. O. (1981). Solar interior structure and luminosity variations, *Sol. Phys.*, 74, pp. 21–34.

Gumsley, A. P., Chamberlain, K. R., Bleeker, W., Söderlund, U., de Kock, M. O., Larsson, E. R. and Bekker, A. (2017). Timing and tempo of the Great Oxidation Event, *Proc. Natl. Acad. Sci. U.S.A.*, 114, pp. 1811–1816.

Harada, M., Tajika, E. and Sekine, Y. (2015). Transition to an oxygen-rich atmosphere with an extensive overshoot triggered by the Paleoproterozoic snowball Earth, *Earth Planet. Sci. Lett.*, 419, pp. 178–186.

Helbig, M., Chasmer, L. E., Kljun, N., Quinton, W. L., Treat, C. C. and Sonnentag, O. (2017). The positive net radiative greenhouse gas forcing of increasing methane emissions from a thawing boreal forest-wetland landscape. *Glob. Change Biol.*, 23, pp. 2413–2427.

Hohmann-Marriott, M. F. and Blankenship, R. E. (2011). Evolution of photosynthesis, *Annu. Rev. Plant Biol.*, 62, pp. 515–548.

Kiang, N. Y. (2008). The color of plants on other worlds, *Sci. Am.*, 298, pp. 48–55.

Kiang, N. Y., Segura, A., Tinetti, G., Govindjee, Blankenship, R. E., Cohen, M., Siefert, J., Crisp, D. and Meadows, V. S. (2007). Spectral signatures of photosynthesis. II. Coevolution with other stars and the atmostphere on extra-solar worlds, *Astrobiology*, 7, pp. 252–274.

Konhauser, K. O., Pecoits, E., Lalonde, S. V., Papineau, D., Nisbet, E. G., Barley, M. E., Arndt, N. T., Zahnle, K. and Kamber, B. S. (2009). Oceanic nickel depletion and a methanogen famine before the Great Oxidation Event, *Nature*, 458, pp. 750–753.

Konhauser, K. O., Robbins, L. J., Pecoits, E., Peacock, C., Kappler, A. and Lalonde, S. V. (2015). The Archean nickel famine revisited, *Astrobiology*, 15, pp. 804–815.

Konn, C., Charlou, J. L., Holm, N. G. and Mousis, O. (2015). The production of methane, hydrogen, and organic compounds in ultramafic-hosted hydrothermal vents of the Mid-Atlantic Ridge, *Astrobiology*, 15, pp. 381–399.

Kothavala, Z., Oglesby, R. J. and Saltzman, B. (1999). Sensitivity of equilibrium surface temperature of CCM3 to systematic changes in atmospheric CO_2, *Geophys. Res. Lett.*, 26, pp. 209–212.

Krasinsky, G. A. and Brumberg, V. A. (2004). Secular increase of astronomical unit from analysis of the major planet motions, and its interpretation, *Celestial Mech. Dyn. Astron.*, 90, pp. 267–288.

Kump, L. P. (2008). The rise of atmospheric oxygen, *Nature*, 451, pp. 277–278.

Lambeck, K. (1975). Effects of tidal dissipation in the oceans on the Moon's orbit and the Earth's Rotation, *J. Geophys. Res.*, 80, pp. 2917–2925.

Lazar, C., McCollom, T. M. and Manning, C. E. (2012). Abiogenic methanogenesis during experimental komatiite serpentinization: Implications for the evolution of the early Precambrian atmosphere, *Chem. Geol.*, 326, pp. 102–112.

Lenton, T. M. and Daines, S. J. (2017). Matworld – The biogeochemical effects of early life on land, *New Phytol.*, 215, pp. 531–537.

Lenton, T. M., Crouch, M., Johnson, M., Pires, N., Dolan, L. (2012). First plants cooled the Ordovician, *Nature Geosci.*, 5, pp. 86–89.

LoPresto, M. C., Hagoort, N. (2011). Determining planetary temperatures with the Stefan-Boltzmann Law, *Physics Teacher*, 49, pp. 113–116.

Luo, G., Ono, S., Beukes, N. J., Wang, D. T., Xie, S. and Summons, R. E. (2016). Rapid oxygenation of Earth's atmosphere 2.33 billion years ago, *Sci. Adv.*, 2, p. e1600134.

Lyons, T. W., Reinhard, C. T. and Planavsky, N. J. (2014). The rise of oxygen in Earth's early ocean and atmosphere, *Nature*, 506, pp. 307–315.

Mestdagh, T. M, Poort, J. and De Batist, M. (2017). The sensitivity of gas hydrate reservoirs to climate change: Perspectives from a new combined model for permafrost-related and marine settings, *Earth-Sci. Rev.*, 169, pp 104–131.

119

Mills, D. B., Ward, L. M., Jones, C., Sweeten, B., Forth, M., Treusch, A. H. and Canfield, D. E. (2014). Oxygen requirements of the earliest animals, *Proc. Natl. Acad. Sci. U.S.A.*, 111, pp. 4168–4172.

Nisbet, E. and Fowlder, C. M. R. (2011). The evolution of the atmosphere in the Archaean and early Proterozoic, *Chin. Sci. Bull.*, 56, pp. 4–13.

Olson, J. M. and Blankenship, R. E. (2004). Thinking about the evolution of photosynthesis, *Photosynth. Res.*, 80, pp. 373–386.

Pavlov, A. A., Hurtgen, M. T., Kasting, J. F. and Arthur, M. A. (2003). Methane-rich Proterozoic atmosphere? *Geology*, 31, pp. 87–90.

Payne, J., McClain, C., Boyer, A. J., Brown, J. H., Finnegan, S., Kowalewski, M., Krause, R., Lyons, S., McShea, D., Novack-Gottshall, P., Smith, F., Spaeth, P., Stempien, J. and Wang, S. (2011). The evolutionary consequences of oxygenic photosynthesis: A body size perspective, *Photosynth. Res.*, 107, pp. 37–57.

Peschek, G. A., Bernroitner, M., Sari, S., Pairer, M. and Obinger, C. (2011). Life inplies work: A holistic account of our microbial biosphere focussing on the bioenergetic processes of cyanobacteria, the ecologically most successful organism on our Earth. *In* Peschek, G. A., Obinger, C., Renger, G., eds, *Bioenergetic Processes of Cyanobacteria: From Evolutionary Singularity to Ecological Diversity* (Springer, Dordrecht), pp. 3–70.

Planavsky, N. J., Asael, D., Hofmann, A., Reinhard, C. T., Lalonde, S. V., Knudsen, A., Wang, X., Ossa Ossa, F., Pecoits, E., Smith, A. J. B., Beukes, N. J., Bekker, A., Johnson, T. M., Konhauser, K. O., Lyons, T. W. and Rouxel, O. J. (2014). Evidence for oxygenic photosynthesis half a billion years before the Great Oxidation Event, *Nat. Geosci.*, 7, pp. 283–286.

Pollack, H. N., Hurter, S. J. and Johnson, J. R. (1993). Heat flow from the Earth's interior: Analysis of the global data set, *Rev. Geophys.*, 31, pp. 267–280.

Renger, G. (1983). Biological energy conservation. *In* Hoppe, W., Lohmann, W., Markl, H., Ziegler, H., eds, *Biophysics* (Springer, Berlin), pp. 347–371.

Ribas, I., Guinan, E. F., Güdel, M. and Audard, M. (2005). Evolution of the solar activity over time and effects on planetary atmospheres. I. High-energy irradiances (1–1700 Å), *Astrophys. J.*, 622, 680–694.

Ribas, I., Porto de Mello, G. F., Ferreira, L. D., Hébrar, E., Selsis, F., Catalán, S., Garcés, A., do Nascimento Jr., J. D. and de Medeiros, J. R. (2010). Evolution of the solar activity over time and effects on planetary atmospheres. II. κ1 Ceti, an analog of the sun when life arose on Earth, *Astrophys. J.*, 714, pp. 384–395.

Riding, R., Fralick, P. and Liang, L. (2014). Identification of an Archean marine oxygen oasis, *Precamb. Res.*, 251, pp. 232–237.

Roberson, A. L., Roadt, J., Halevy, I. and Kasting, J. F. (2011). Greenhouse warming by nitrous oxide and methane in the Proterozoic eon, *Geobiology*, 9, pp. 313–320.

Rosing, M. T. and Frei, R. (2004). U-rich Archaean sea-floor sediments from Greenland – Indications of >3700 Ma oxygenic photosynthesis, *Earth Planet. Sci. Lett.*, 217, pp. 237–244.

Ruppel, C. D. and Kessler, J. D. (2017). The interaction of climate change and methane hydrates, *Rev. Geophys.*, 55, pp. 126–168.

Satkoski, A. M., Beukes, N. J., Li, W., Beard, B. L. and Johnson, C. M. (2015). A redox-stratified ocean 3.2 billion years ago, *Earth Planet. Sci. Lett.*, 430, pp. 43–53.

Schirrmeister, B. E., Gugger, M. and Donoghue, P. C. J. (2015). Cyanobacteria and the Great Oxidation Event: Evidence from genes and fossils, *Palaeontology*, 58, pp. 769–785.

Schopf, J. W. (2011). The paleobiological record of photosynthesis, *Photosynth. Res.*, 107, pp. 87–101.

Schwarzschild, B. M. (2014). Neutrinos from Earth's interior measure the planet's radiogenic heating, *Phys. Today*, 64, pp. 14–17.

Segura, A., Krelove, K., Kasting, J. F., Sommerlatt, D., Meadows, V., Crisp, D., Cohen, M. and Mlawer, E. (2003). Ozone concentrations and ultraviolet fluxes on Earth-like planets around other stars, *Astrobiology*, 3, pp. 689–708.

Sleep, N. H. and Bird, D. K. (2008). Evolutionary ecology during the rise of dioxygen in the Earth's atmosphere, *Philos. Trans. R. Soc. Lond.*, B, 363, pp. 2651–2664.

Shevela, D., Pishchalnikov, R. Y., Eichacker, L. A. and Govindjee (2013). Oxygenic photosynthesis in cyanobacteria. *In* Srivastava, A. K., Amar, N. R., Neilan, B. A., eds, *Stress Biology of Cyanobacteria: Molecular Mechanisms to Cellular Responses* (CRC Press/Taylor & Francis Group, Boca Raton, FL), pp. 3–40.

Tashiro, T., Ishida, A., Hori, M., Igisu, M., Koike, M., Méjcan, P., Naoto Takahata, N., Yuji Sano, Y. and Komiya, T. (2017). Early trace of life from 3.95 Ga sedimentary rocks in Labrador, Canada, *Nature*, 549, pp. 516 519.

Tajika, E. (2003). Faint young Sun and the carbon cycle: Implication for the Proterozoic global glaciations, *Earth Planet. Sci. Lett.*, 214, pp. 443 453

Thomassot, E., O'Neil, J., Francis, D., Cartigny, P. and Wing, B. A. (2015). Atmospheric record in the Hadean Eon from multiple sulfur isotope measurements in Nuvvuagittuq Greenstone Belt (Nunavik, Quebec), *Proc. Natl. Acad. Sci. U.S.A.*, 112, pp. 707–712.

Ward, L. M., Kirschvink, J. L. and Fischer, W. W. (2016). Timescales of oxygenation following the evolution of oxygenic photosynthesis, *Orig. Life Evol. Biosph.*, 46, pp. 51–65.

Woese, C. R., Kandler, O. and Wheelis, R. M. (1990). Towards a natural system of organisms: Proposal for the domains Archaea, Bacteria, and Eucarya, *Proc. Natl Acad. Sci. U.S.A.*, 87, pp. 4576–4579.

Wongnate, T., Sliwa, D., Ginovska, B., Smith, D., Wolf, M. W., Lehnert, N., Raugei, S. and Ragsdale, S. W. (2016). The radical mechanism of biological methane synthesis by methylcoenzyme M reductase, *Science*, 352, pp. 953–958.

Zellner, E. B. (2017). Cataclysm no more: New views on the timing and delivery of lunar impactors, *Orig. Life Evol. Biosph.*, 47, pp. 261–280.

Zhang, W. J., Li, Z. B. and Lei, Y. (2010). Experimental measurement of growth patterns on fossil corals: Secular variation in ancient Earth-Sun distances, *Chin. Sci. Bull.*, 55, pp. 4010–4017.

Chapter 7
Anoxygenic Photosynthesis

7.1 Introduction

Anoxygenic photosynthesis, i.e., photosynthesis without evolution of oxygen, is carried out by a large number of photosynthetic bacteria from different phylogenetic groups. Most of them are photoautotrophic, and use inorganic carbon as their carbon source, while others, e.g., some purple non-sulfur bacteria are photoheterotrophic, and use organic compounds. Raven [2009] estimates that chemolithotrophs and anoxygenic photolithotrophs together contribute about 0.40 Pg C/year to the overall Net Primary Productivity (NPP), which is less than a percent of what oxygenic phototrophs provide. For detailed discussion of anoxygenic photosynthesis, see reviews [Blankenship *et al.*, 1995; Hunter *et al.*, 2009].

In oxygenic photosynthesis, water serves as the ultimate electron source, whereas in anoxygenic photosynthesis, electron donors include sulfur compounds, ferrous ions, or molecular hydrogen. Instead of chlorophylls, pigments used in oxygenic photosynthesis, anoxygenic bacteria use bacteriochlorophylls (BChls), whose absorption extends into the infrared (Fig. 7.1a,b) (see a recent review on photosynthetic pigments by Larkum *et al.* [2018]). Absorption spectra of BChls are influenced by their association with other pigments as well as with themselves (i.e., oligomerization), and interaction with specific amino acids; absorption in the infrared region gives them a niche that cannot be exploited by oxygenic photosynthesizers. Most extreme in this respect are some species of *Blastochloris*, *Rhodopseudomonas* and

Photosynthesis: Solar Energy for Life by Dmitry Shevela, Lars Olof Björn and Govindjee
© 2018, published by World Scientific Publishing Co. Pte. Ltd. ISBN: 978-981-3223-10-3.

(a)

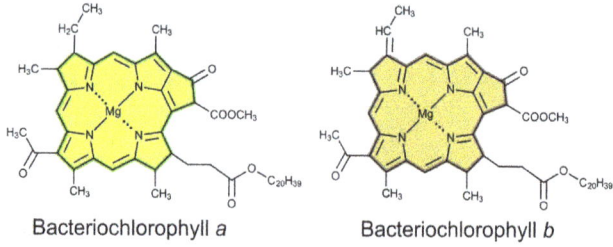

Bacteriochlorophyll *a*   Bacteriochlorophyll *b*

(b)

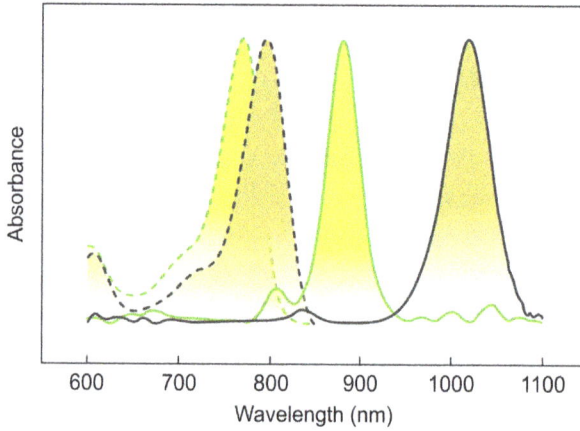

(c)

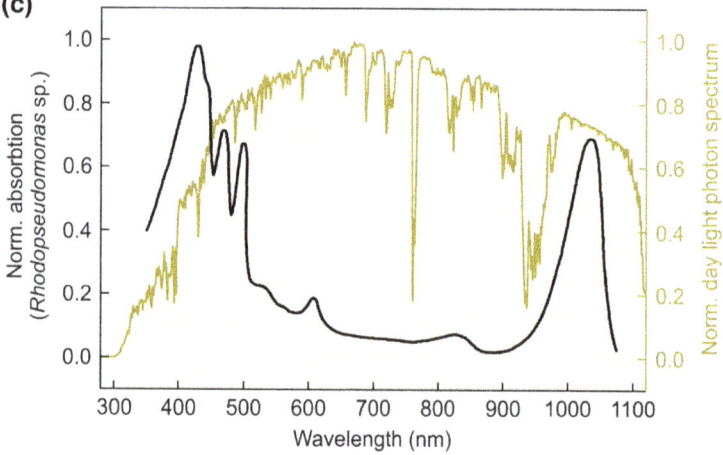

(continued)

Fig. 7.1. (continued) Bacteriochlorophylls (BChls) and their absorption spectra. **(a)** Structural formula of BChl *a* and BChl *b*. **(b)** Absorption spectra of BChl *a* and BChl *b* (in color; cf. panel a) in solution (dashed curves) and in reaction center-light harvesting complex 1 (RC-LH1) (solid light green line is for, *Rhodobacter sphaeroides* and solid brown line is for *Blastochloris viridis*). Modified from Canniffe and Hunter [2014]. **(c)** Daylight spectrum (brownish yellow curve), calculated from ASTM G173-03 reference spectrum for global radiation on a tilted surface derived from SMARTS v. 2.9.2, and absorption spectrum of a suspension of *Rhodopseudomonas* sp. NHTC 133 (black curve; redrawn from Olson and Nadler, 1965); norm. stands for normalized, maximum is taken as 1.

Ectothiorhodospira containing BChl *b* [Olson and Nadler, 1965; Imhoff and Trüper, 1977; Canniffe and Hunter, 2014]. These bacterial species have absorption bands beyond 1000 nm (1 μm), and, thus, at longer wavelength than the pigments of all other competitors for light, and, most importantly, also at wavelengths longer than the strong absorption of atmospheric water vapor, which is between 900 nm and 1000 nm (Fig. 7.1c).

The relation between the absorption spectrum of water and of photosynthetic antenna pigments has been discussed by Stomp *et al.* [2007]. For different pathways for the synthesis of BChl *b*, see [Tsukatani *et al.*, 2013], and for a complete discussion of all aspects of these and related pigments, see [Grimm *et al.*, 2006].

7.2 Anoxygenic Photosynthetic Organisms: Their Reaction Centers and Pathways for Carbon Assimilation

Two different reaction centers (RCs) in two different photosystems, PSI and PSII, are involved in oxygenic photosynthesis; their excitation energy "traps" have been labeled as P700 and P680 (see Chapters 2 and 3). In anoxygenic photosynthesis, there is always only one kind of RC involved, but it is related to either a PSI or PSII RC, depending on the bacteria. We designate these bacterial RCs to be of either type I (type 1) or type II (type 2). Many anoxygenic photosynthetic bacteria use, for the assimilation of carbon, the same Calvin-Benson cycle as cyanobacteria and eukaryotes do (Fig. 7.2). However, there are also other pathways, several of which are shown in Fig. 7.3. Of these the reductive

16/18S rRNA		Reaction Centers		Fixation/Assimilation Pathway			
				CO$_2$			Acetate
		RCI	RCII	CBB	rTCA	3-HPP	EtMa-CoA
Archaea		−	−	−	+	+	−
Chloroflexi		−	+	−	−	+	−
Alphaproteobacteria		−	+	+	−	−	+
Gammaproteobacteria		−	+	+	−	−	−
Heliobacteria		+	−	−	Partial	−	−
Chlorobia		+	−	−	+	−	−
Cyanobacteria/Prochlorophytes		+	+	+	−	−	−
Eukarya		+	+	+	−	−	−

Tree node values: 84, 98, 90, 84, 99, 99, 100, 52, 47, 100, 54, 86, 100

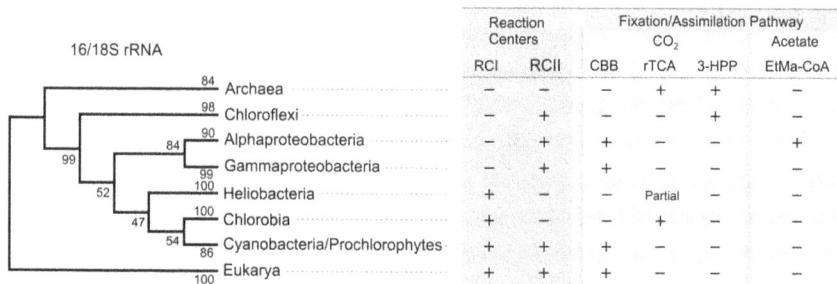

Fig. 7.2. Overview of prokaryotic phototrophs, with Archaea and Eukarya included for reference. Redrawn from Hanson *et al.* [2012]. Chloroflexi (plural for Chloroflexus) are nonsulfur green bacteria. For details of the fixation/assimilation pathways, see Fig. 7.3.

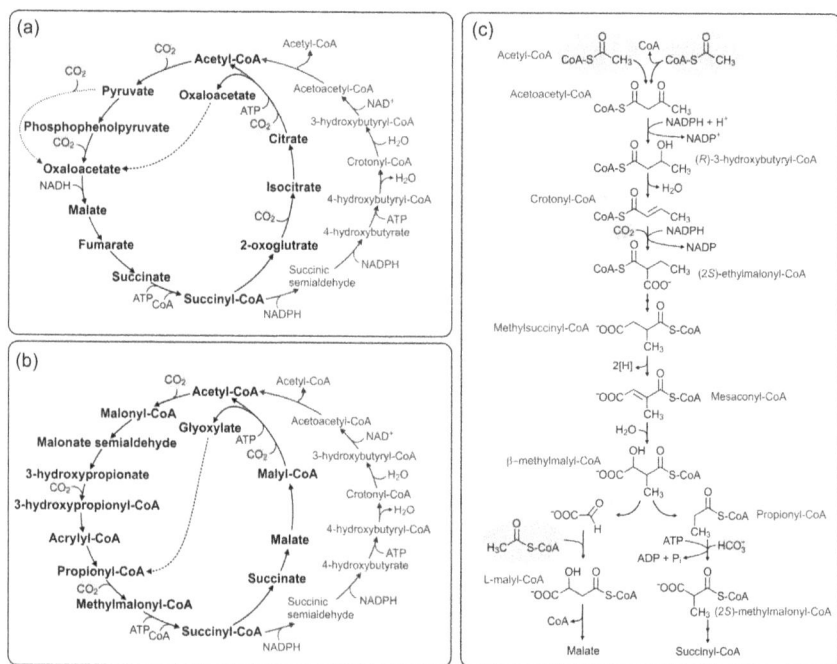

Fig. 7.3. Carbon assimilation pathways. **(a)** The reductive tricarboxylic acid cycle (rTCA) is shown in black, whereas, the steps specific for the dicarboxylate/4-hydroxybutyrate cycle, operating only in archaea (non-phototrophic organisms), are in grey; the steps from acetyl-CoA to succinyl-CoA are the same as for the rTCA. **(b)** The 3-hydroxypropionate cycle (3-HPP) is shown in black, whereas the steps specific for the 3-hydroxypropionate/4-hydroxy-butyrate cycle, operating only in (non-phototrophic) archaea, are in grey. The steps from acetyl-CoA to succinyl-CoA are common to both cycles. The panels (a) and (b) are modified from Saini *et al.* [2011]; the right part **(c)** is the ethylmalonyl CoA pathway for the assimilation of acetyl-CoA, CO$_2$, and bicarbonate. Modified from Berg [2011] and Hanson *et al.* [2012].

tricarboxylic acid pathway was discovered in *Chlorobium*, a green sulfur bacterium [Evans *et al.*, 1966]. For the Arnon-Buchanan cycle in several photosynthetic bacteria, see [Buchanan *et al.*, 2017].

7.3 Relation between the RCs of Anoxygenic and Oxygenic Photosynthetic Organisms

Since the type I and type II RCs are very similar to each other, it is very likely that they had evolved from a common ancestral protein (Fig. 7.4). Since cyanobacteria have both types of RCs, the question arises as to how

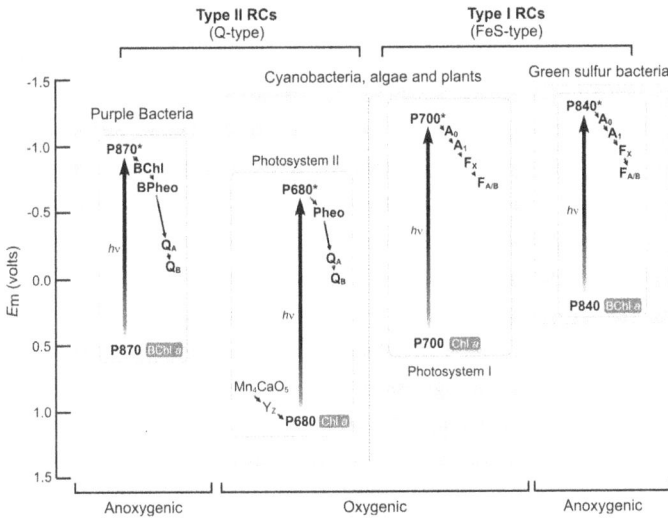

Fig. 7.4. Type II (or type 2) and type I (or type 1) reaction centers (RCs) in anoxygenic (purple bacteria and green sulfur bacteria) and oxygenic (cyanobacteria, algae, and plants) organisms. The electron donors of the RCs are presented according to their midpoint redox potential values (E_m, at pH 7). Depending on the final electron acceptor (quinone [Q_A/Q_B] or iron-sulfur clusters [F_X, F_A, and F_B]), the RCs are classified as Q- or FeS-type. The number after P indicates the wavelength of one of the absorption maxima of the RC pigment, and is inversely related to the absorbed photon energy (note the various lengths of straight black arrows). Some redox components of the electron transport chain are omitted for the sake of simplicity. A_0 is an electron acceptor, which represents BChl in green sulfur bacteria or Chl in PSI of oxygenic organisms. A_1 is an electron acceptor, which is a phylloquinone in PSI RC of oxygenic organisms, but its identity in green sulfur bacteria is still uncertain. (Note: BPheo is for bacteriopheophytin.) Based on the data in Hohmann-Marriott and Blankenship [2011] and Blankenship [2014]. Modified from Shevela *et al.* [2013].

127

this has come about. Earlier, it was thought that there has been either a merger between bacteria possessing type I and type II, or there has been a massive horizontal gene transfer. John Allen [2005] has, however, suggested that the differentiation between type I and type II RCs may have taken place in the same organism, which gave rise to cyanobacteria, and then the anoxygenic single-RC photosynthesis systems arose by the loss of one of the two photosystems.

7.4 Photosynthetic N_2 Assimilation

Many anaerobic photosynthetic bacteria, capable of photoassimilation of CO_2, are also able to assimilate molecular nitrogen, N_2. Some, such as *Rhodopseudomonas capsulata*, *Rhodopseudomonas acidophila* [Meyer *et al.*, 1978; Siefert and Pfennig, 1980] and *Thiocapsa* sp. strain 5811 [Jouanneau *et al.*, 1980], are able to assimilate N_2 even with O_2 present at low levels, despite the sensitivity of nitrogenase to O_2. The greater demand for ATP (i.e., 18 ATP per N_2 assimilated), in these systems, is partly satisfied by photophosphorylation, while the electrons (8 to 9 per N_2) for reduction to the level of ammonia come primarily from carbon assimilates, *via* glycolysis or the oxidative pentose phosphate pathway. Stoichiometrically, only six electrons would be needed for reducing one N_2 to ammonia, but some electrons are used for the release of H_2. Examples of bacteria, with anoxygenic nitrogen fixing phototrophs, within different phyla are:

- Alphaproteobacteria: Purple nonsulfur bacteria (*Rhodobacter*, *Rhodomicrobium*, *Rhodospirillum*, *Rhodopseudomonas*).
- Gammaproteobacteria: Purple sulfur bacteria (*Chromatiaceae*: *Amoebobacter*, *Chromatium*, *Thiocapsa*, *Thiocystis*) and *Ectothiorhodospiraceae* (*Ectothiorhodospira*).
- Green sulfur bacteria: Chlorobi (*Chlorobium*, *Pelodictyon*, *Prosthecochloris*).
- Firmicutes: *Heliobacteriaceae* (*Heliobacterium*).

Most cyanobacteria are able to assimilate N_2, in spite of living in an oxygen-rich environment, and even themselves producing oxygen. They have two strategies for managing this difficult task. Some, especially the single-celled forms, release O_2 in the daytime, and assimilate N_2 during the night [Bergman *et al.*, 1997; Falcón *et al.*, 2004]. But already long ago, some cyanobacteria had developed a sophisticated method, which allowed them to assimilate N_2 under aerobic conditions and in full daylight, by spatial separation of oxygen release and nitrogen assimilation, although other solutions to the oxygen sensitivity of nitrogenase have evolved [Bergman *et al.*, 1997; 2013]. These cyanobacteria form filaments consisting of rows of "vegetative" cells carrying out oxygenic photosynthesis, here and there interspersed with so-called *heterocysts*. Nitrogen assimilation occurs in these specialized cells that are surrounded by a thick, oxygen-excluding envelope. These cells have a thylakoid system with PSI, but they lack PSII, and thus they cannot produce O_2. How the necessary electrons and ATP are supplied to the nitrogenase in heterocysts is still debated, but two possibilities are shown in Fig. 7.5. In both cases, electrons are supplied *via* NADPH from the oxidative pentose pathway, OPP, and glucose-6-phosphate is supplied from the vegetative cells. In one model (Fig. 7.5a), cyclic electron transport, driven by PSI, produced ATP. However, Almon and Böhme [1982] as well as Magnuson and Cardona [2016] (Fig. 7.5b) showed that PSI drives non-cyclic electron transport from NADPH to nitrogenase, and, that this electron transport, coupled with proton transport across the thylakoid membrane, also leads to the synthesis of ATP.

The thick envelope around heterocysts consists of several layers (Fig. 7.6). As in other cells, the cytoplasm is surrounded by a cell membrane. Further, both the heterocysts, and the vegetative cyanobacterial cells, have periplasm as well as an outer membrane; heterocysts have not only a glycolipid layer, but also a very thick polysaccharide layer. Murry and Wolk [1989] showed that both the heterocyst-specific layers are needed for excluding oxygen from the heterocyst and allowing nitrogenase to function under aerobic environments.

129

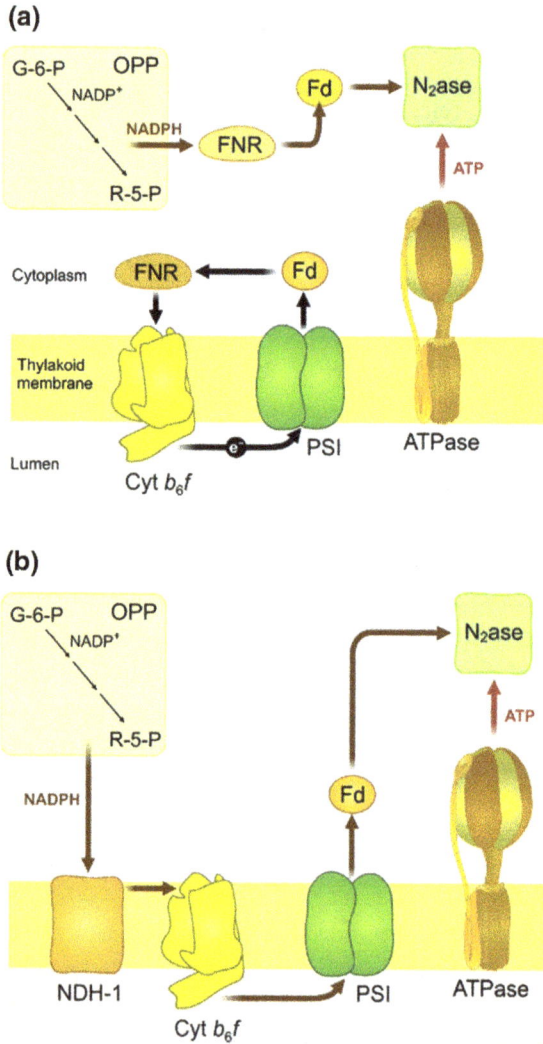

Fig. 7.5. Models for a possible supply of the electrons and ATP to nitrogenase (N₂ase) in the heterocysts. In both models (a and b), shown here, electrons are supplied *via* NADPH from the oxidative pentose pathway (OPP), and glucose-6-phosphate is supplied from the vegetative cells. In an earlier model **(a)** ATP-producing cyclic electron transport was driven by PSI. However, in the model shown in **(b)**, PSI also drives non-cyclic (linear) electron transport. For further explanation, see the main text. Modified from Magnuson and Cardona [2016].

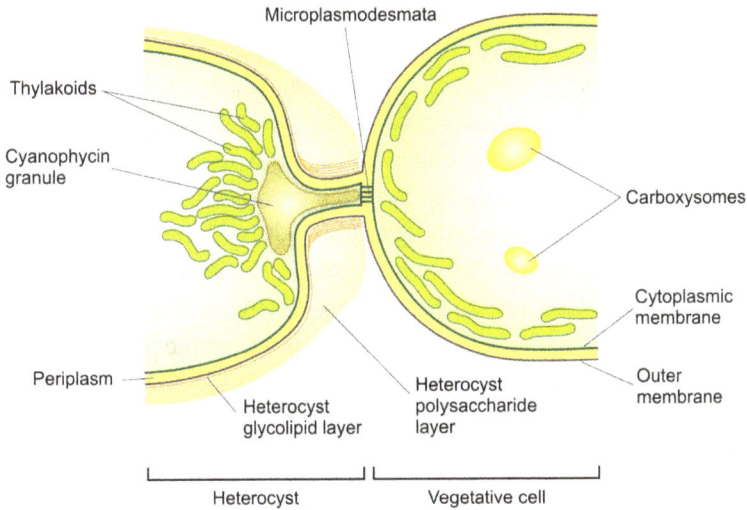

Fig. 7.6. Schematic representation of a heterocyst, which is in association with a vegetative cell of a filamentous cyanobacterium. Note that the layers surrounding the interior of both cell types are a cell membrane, periplasm, and an outer membrane. The heterocyst has, in addition, glycolipid and polysaccharide layers. Modified and redrawn from Flores and Herrero [2010].

7.5 Rhodopsin-Based Phototrophy

All the organisms dealt with thus far carry out photosynthesis based on chlorophyll-type pigments (Chls or BChls). But there exists also a completely different way of utilizing the energy from the sun, based on pigments related to our visual pigments, the rhodopsins. Prokaryotic rhodopsins have a similar structure as our rhodopsins, but have completely different amino acid sequences, and the evolutionary relationship between the two kinds of rhodopsin is still being debated. In many cases, bacterial rhodopsins serve as sensors for phototaxis, or as light powered ion pumps. In case the ions transported across a bacterial membrane are protons, the transport can be linked to the synthesis of ATP via an ATP synthase (ATPase) in the same membrane, just as in other kinds of photophosphorylation. For information on the quantum efficiency of proton pumping in this system, see R. Govindjee *et al.* [1990].

131

There is, in particular, one kind of bacterial rhodopsin, i.e., proteorhodopsin, which plays a large role in light-driven proton pump [Beja *et al.*, 2001], as it has been found in many organisms, even in some eukaryotes [Slamovits *et al.*, 2011], and Archaea [Frigaard *et al.*, 2006] *via* lateral (horizontal) gene transfer, probably in part through viruses [Yutin and Koonin, 2012]. No case is known in which proteorhodopsin is involved in assimilation of inorganic carbon, but light-induced ion pumping and ATP generation helps to conserve organic carbon [Courties *et al.*, 2015] and promote survival [Gómez-Consarnau *et al.*, 2010].

In Chapter 8 we shall discuss photosynthesis from the perspective of the past, present and the future.

References

Allen, J. F. (2005). A redox switch hypothesis for the origin of two light reactions in photosynthesis, *FEBS Lett.*, 579, pp. 963–968.

Almon, H. and Böhme, H. (1982). Photophosphorylation in isolated chloroplasts from the blue-green alga *Nostoc muscorum*, *Biochim. Biophys. Acta*, 679, pp. 279–296.

Beja, O., Spudich, E. N., Spudich, J. L., Leclerc, M. and DeLong, E. F. (2001). Proteorhodopsin phototrophy in the ocean, *Nature*, 411, pp. 786–789.

Berg, I. A. (2011). Ecological aspects of the distribution of different autotrophic CO_2 fixation pathways, *Appl. Environ. Microbiol.*, 77, pp. 1925–1936.

Bergman, B., Gallon, J. R., Rai, A. N. and Stal, L. J. (1997). N_2 fixation by non-heterocystous cyanobacteria, *FEMS Microbiol. Rev.*, 19, pp. 139–185.

Bergman, B., Sandh, G., Lin, S., Larsson, J. and Carpenter, E. J. (2013). Trichodesmium – A widespread marine cyanobacterium with unusual nitrogen fixation properties, *FEMS Microbiol. Rev.*, 37, pp. 286–302.

Blankenship, R. E. (2014). *Molecular Mechanisms of Photosynthesis*. 2nd ed. (Wiley Blackwell, Oxford).

Blankenship, R. E., Madigan, M. T. and Bauer, C. E., eds., (1995). *Anoxygenic Photosynthetic Bacteria* (Kluwer Academic Publishers, Dordrecht).

Buchanan, B. B., Sirevåg, R, Fuchs, G., Ivanovsky, R. N., Igarashi, Y., Ishii, M., Tabita, F. R. and Berg, I. A. (2017). The Arnon–Buchanan cycle: A retrospective, 1966–2016, *Photosynth. Res.*, 134, pp.117–131.

Canniffe, D. P. and Hunter, C. N. (2014). Engineered biosynthesis of bacteriochlorophyll *b* in *Rhodobacter sphaeroides*, *Biochim. Biophys. Acta*, 1837, pp. 1611–1616.

Courties, A., Riedel, T., Rapaport, A., Lebaron, P. and Suzuki, M. T. (2015). Light-driven increase in carbon yield is linked to maintenance in the proteorhodopsin-containing *Photobacterium angustum* S14, *Front. Microbiol.*, 6, p. 688.

Evans, M. C., Buchanan, B. B. and Arnon, D. I. (1966). A new ferredoxin-dependent carbon reduction cycle in a photosynthetic bacterium, *Proc. Natl. Acad. Sci. U.S.A.*, 55, pp. 928–934.

Falcón, L. I., Carpenter, E. J., Cipriano, F., Bergman, B. and Capone, D. G. (2004). N_2 fixation by unicellular bacterioplankton from the Atlantic and Pacific Oceans: Phylogeny and *in situ* rates, *Appl. Environ. Microbiol.*, 70, pp. 765–770.

Flores, E. and Herrero, A. (2010). Compartmentalized function through cell differentiation in filamentous cyanobacteria, *Nat. Rev. Microbiol.*, 8, pp. 39–50.

Frigaard, N.-U., Martinez, A., Mincer, T. J. and DeLong, E. F. (2006). Proteorhodopsin lateral gene transfer between marine planktonic bacteria and archaea, *Nature*, 439, pp. 847–850.

Gómez-Consarnau, L., Akram, N., Lindell, K., Pedersen, A., Neutze, R., Milton, D. L., González, J. M. and Pinhassi, J. (2010). Proteorhodopsin phototrophy promotes survival of marine bacteria during starvation, *PLOS Biol.*, 8, p. e1000358.

Govindjee, R., Balashov, S. and Ebrey, T, (1990). Quantum efficiency of the photochemical cycle of bacteriorhodopsin, *Biophys. J.*, 58, pp. 597–608.

Grimm, B., Porra, R. J., Rudiger, W. and Scheer, H., eds., (2006). *Chlorophylls and Bacteriochlorophylls: Biochemistry, Biophysics, Functions and Applications* (Springer, Dordrecht).

Hanson, T. E., Alber, B. E., Tabita, F. R. (2012). Phototrophic CO_2 fixation: Recent insights into ancient metabolisms. *In* Burnap, R. L. and Vermaas, W. F. J., eds, *Functional Genomics and Evolution of Photosynthetic Systems* (Springer, Dordrecht), pp. 225–251.

Hohmann-Marriott, M. F. and Blankenship, R. E. (2011). Evolution of photosynthesis, *Annu. Rev. Plant Biol.*, 62, pp. 515–548.

Hunter, C. N., Daldal, F., Thurnauer, M. C. and Beatty, J. T., eds., (2009). *The Purple Phototrophic Bacteria*, Vol. 28 (Springer, Dordrecht).

Imhoff, J. F. and Trüper, H. G. (1977). *Ectothiorhodospira halochloris* sp. nov., a new extremely halophilic phototrophic bacterium containing bacteriochlorophyll *b*, *Arch. Microbiol.*, 114, pp. 115–121.

Jouanneau, Y., Siefert, E. and Pfennig, N. (1980). Microaerobic nitrogenase activity in *Thiocapsa* sp. strain 5811, *FEMS Microbiol. Lett.*, 9, pp. 89–93.

Larkum, A. W. D., Ritchie, R. J. and Raven, J. A. (2018). Living off the Sun: Chlorophylls, bacteriochlorophylls and rhodopsins, *Photosynthetica*, 56, pp. 11–43.

Magnuson, A. and Cardona, T. (2016). Thylakoid membrane function in heterocysts, *Biochim. Biophys. Acta*, 1857, pp. 309–319.

Meyer, J., Kelley, B. C. and Vignais, P. M. (1978). Nitrogen-fixation and hydrogen metabolism in photosynthetic bacteria, *Biochimie*, 60, pp. 245–260.

Murry, M. A. and Wolk, C. P. (1989). Evidence that the barrier to the penetration of oxygen into heterocysts depends upon two layers of the cell envelope, *Arch. Microbiol.*, 151, pp. 469–474.

Olson, J. M. and Nadler, K. D. (1965). Energy transfer and cytochrome function in a new type of photosynthetic bacterium, *Photochem. Photobiol.*, 4, pp. 891–900.

Raven, J. A. (2009). Contributions of anoxygenic and oxygenic phototrophy and chemolithotrophy to carbon and oxygen fluxes in aquatic environments, *Aquat. Microb. Ecol.*, 56, pp. 177–192.

Saini, R., Kapoor, R., Kumar, R., Siddiqi, T. O. and Kumar, A. (2011). CO_2 utilizing microbes – A comprehensive review, *Biotechnol. Adv.*, 29, pp. 949–960.

Siefert, E. and Pfennig, N. (1980). Diazotrophic growth of *Rhodopseudomonas acidophila* and *Rhodopseudomonas capsulata* under microaerobic conditions in the dark, *Arch. Microbiol.*, 125, pp. 73–77.

Shevela, D., Pishchalnikov, R. Y., Eichacker, L. A. and Govindjee (2013). Oxygenic photosynthesis in cyanobacteria. *In* Srivastava, A. K., Amar, N. R. and Neilan, B. A., eds, *Stress Biology of Cyanobacteria: Molecular Mechanisms to Cellular Responses* (CRC Press/Taylor & Francis Group, Boca Raton, FL), pp. 3–40.

Slamovits, C. H., Okamoto, N., Burri, L., James, E. R. and Keeling, P. J. (2011). A bacterial proteorhodopsin proton pump in marine eukaryotes, *Nat. Commun.*, 2, p. 183.

Stomp, M., Huisman, J., Stal, L. J. and Matthijs, H. C. P. (2007). Colorful niches of phototrophic microorganisms shaped by vibrations of the water molecule, *ISME J.*, 1, pp. 271–282.

Tsukatani, Y., Yamamoto, H., Harada, J., Yoshitomi, T., Nomata, J., Kasahara, M., Mizoguchi, T., Fujita, Y. and Tamiaki, H. (2013). An unexpectedly branched biosynthetic pathway for bacteriochlorophyll *b* capable of absorbing near-infrared light, *Sci. Rep.*, 3, p. 1217.

Yutin, N. and Koonin, E. V. (2012). Proteorhodopsin genes in giant viruses, *Biology Direct*, 7, pp. 34–34.

Chapter 8

The Past, Present and the Future

8.1 Spread of Photosynthesis by Successive Endosymbiosis

As we have already mentioned, oxygenic photosynthesis first appeared among the cyanobacteria. It then "spread" to other organisms by a process called *endosymbiosis*: the uptake and inclusion of a cyanobacterium into another organism. We do not have any information on this "host organism"; in all likelihood, it is extinct. With the exception of a few examples, to be discussed later, all eukaryotic phototrophic organisms (plants and algae) may have evolved from a single endosymbiotic event. When precisely this event took place is not known, and estimates vary from 900 Ma ago [Shih and Matzke, 2013] to around 2,000 Ma ago [Blank, 2013]; also see Yoon *et al.* [2004] who place it somewhere in between. According to one theory, *Chlamydia*-like bacteria were also involved in this event [Ball *et al.*, 2013; Ball *et al.*, 2015; Ball *et al.*, 2016]; however, this theory has been refuted by Domman *et al.* [2015]. The possibility of multiple primary endosymbiosis has been discussed by Stiller [2003], Howe *et al.* [2008], and Larkum *et al.* [2007], but in the following discussion, we shall assume it to be a single event. Ochoa de Alda *et al.* [2014] have used rRNA sequences to find which cyanobacterium is closest to the one that participated in the first endosymbiosis event. The fact that chloroplasts group with late diverging cyanobacteria speaks against a very early primary endosymbiotic event. The primary symbiont likely contained both Chl *b* and phycobiliproteins [Tomitani *et al.*, 1999]; further, there

Photosynthesis: Solar Energy for Life by Dmitry Shevela, Lars Olof Björn and Govindjee
© 2018, published by World Scientific Publishing Co. Pte. Ltd. ISBN: 978-981-3223-10-3.

still lives a cyanobacterium (*Prochlorococcus marinus* CCMP 1375) that has both Chl *b* and a phycobiliprotein [Hess *et al.*, 1996].

According to the most commonly held view, the first group to diverge among eukaryotic photosynthesizers was Glaucophyta [Jackson *et al.*, 2015] (Fig. 8.1). Later there was a split between what has become known as the "green line" and the "red line" for further evolution of eukaryotic photosynthesizers. Yoon *et al.* [2004], using plastid genomes, date the divergence of Glaucophyta from the main line to 1,558 Ma ago, and the split between the "green line" and the "red line" to 1,474 Ma ago. Using information from nuclear genes, Parfrey *et al.* [2011] arrived at very similar divergence times. However, Blank [2013] and Ševčíková *et al.* [2015] have presented somewhat different scenarios for the early evolution.

The further evolution of the "green line" is rather straightforward and "tree-like", with the exception of the occasional horizontal transfer of genes between organisms that are only distantly related, and the occasional transfer of chloroplasts to other organisms (e.g., gastropods, [Händeler *et al.*, 2009]; dinoflagellates, [Kamikawa *et al.*, 2015]; euglenids, [Hrdá *et al.*, 2012; Bennett and Triemer, 2015]; chlorarachniophytes, [Suzuki *et al.*, 2015]). The "red line", in contrast, exhibits complications in that the plastids and their host cells have different origins and are, in some cases, reminiscent of Russian *matroshka* dolls with layer upon layer of different origin.

It is important to distinguish between nuclear (Fig. 8.1) and plastid (Fig. 8.2) phylogeny. The plastids on the "red line" are derived from the red algae. At one point, according to Yoon *et al.* [2004], about 1,250 Ma ago, there occurred a secondary endosymbiotic event, in which an organism "engulfed" a red alga and converted it into a chloroplast. Further, there have been several secondary endosymbiotic events involving the uptake of red algae by different hosts. However, it appears certain that there was a single uptake event for apicomplexan, dinoflagellate, and heterokont (stramenopile) plastids [Janouškovec *et al.*, 2010].

The second endosymbiotic event may have been followed by a third (tertiary) and even a fourth (quaternary) endosymbiosis. Dinoflagellates appear to have been exceptional in their ability for endosymbiosis, and they may have incorporated not only cryptomonads, but haptophytes, diatoms,

Stramenopile-alveolate-rhizarian clade

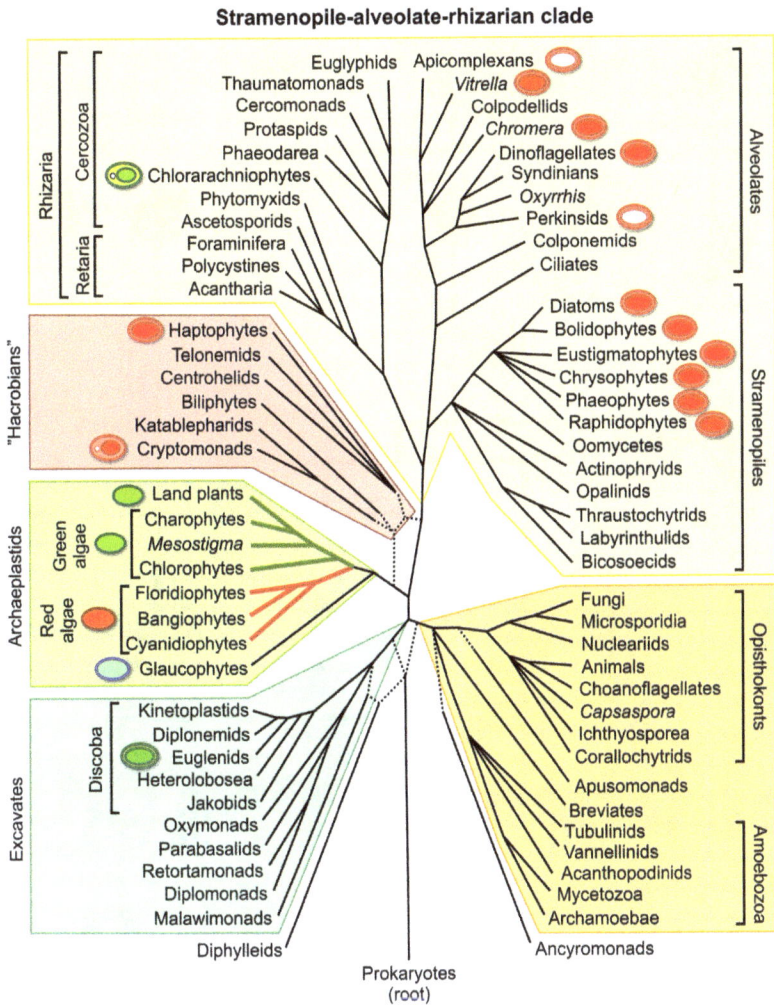

Fig. 8.1. Nuclear phylogeny of plastid-containing organisms. Chloroplasts on the "red line" are indicated in red, and on the "green line" in green; the white indicates that photosynthetic ability had been lost. Uncertain evolutionary paths are dashed. For plastids that arose by secondary endosymbiosis, the nucleomorph (small circle) and periplastic space (lighter green or red color) present in chlorarachniophytes and cryptomonads are indicated. The chloroplast of Glaucophyta, known as 'cyanoplast' is shown in cyan. Modified from Keeling [2013] with permission by Patrick J. Keeling.

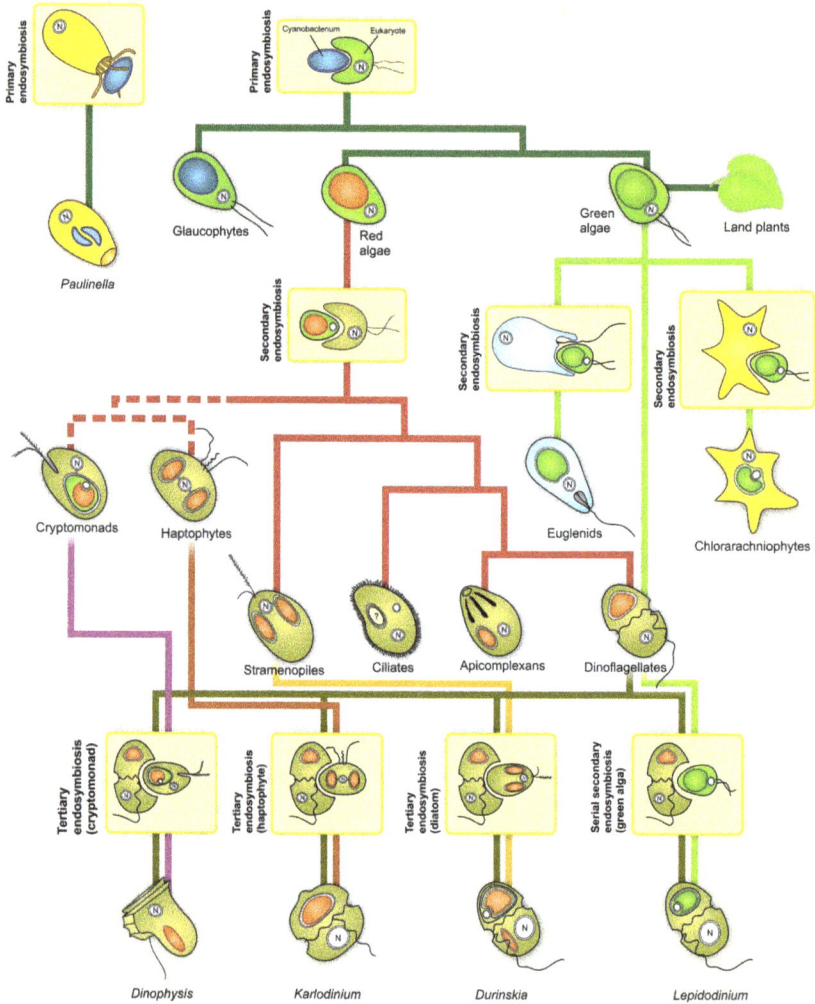

Fig. 8.2. Plastid genesis by primary, secondary, serial secondary, and tertiary endosymbiosis. Modified from Keeling [2013] with permission by Patrick J. Keeling.

or green algae, converting them to be just chloroplasts, as replacement(s) for the original red alga (Fig. 8.2).

The terminology in algal systematics is confusing, due mainly to the existence of several phylogenetic theories. "Stramenopiles" (near the center of Fig. 8.2) are almost synonymous with "heterokonts", but some researchers include the non-photosynthetic oomycetes and hyphochythrids among heterokonts but not among stramenopiles. The heterokonts consist of brown algae, golden algae, and the extremely important diatoms (Bacillariophyta or Diatomae). Diatoms carry out 40% of the marine and 20% of global primary production of organic matter. An obsolete group is Chromoalveolata, which includes everything on the "red line" below the red algae in Fig. 8.2. (We note that the haptophytes (Haptophyta) are often referred to as prymnesiophytes (Prymnesiophyta) by many authors.)

As mentioned above, there are exceptions to the statement that there has been only a single primary endosymbiotic event. The most remarkable exception is the photosynthetic freshwater amoeba *Paulinella chromatophora* [Lauterborn, 1895; Nakayama and Archibald, 2012; Nowack, 2014] of the phylum Cercozoa (Fig. 8.2). It has two chloroplasts per cell of Synechococcus/Prochlorococcus-like cyanobacterial origin. Recently a marine species, *Paulinella longichromatophora*, was found to also possess similar chloroplasts, i.e., two per cell [Kim and Park, 2016].

In all of the above cases, the endosymbiosis has resulted in the ability of the host to take advantage of its capacity to assimilate carbon dioxide photosynthetically. However, in some cases the ability of the acquired plastid to carry out photosynthesis has been lost. The best-known example is the parasitic group Apicomplexa (see Figs. 8.1 and 8.2), of which the most well-known members are the malaria parasites and Toxoplasma. The fact that plastids have been retained in these organisms indicates that they may have other important functions.

Most free-living cyanobacteria are able to assimilate both CO_2 and N_2. The ability to assimilate N_2 was either lost already in the cyanobacterial ancestor of the plastids, or during primary endosymbiosis. A sort of parallel to the endosymbiosis that led to chloroplasts is cyanobacterial symbiosis, in which the cyanobacterial partner (belonging to the unicellular cyanobacteria group A, UCYN-A) lost the ability to assimilate CO_2 but retained the ability to assimilate N_2 [Zehr *et al.*, 2016]. There are cyanobacterial remnants, known as spheroidal bodies inside the diatoms of the genera *Rhopalodia* [Kneip *et al.*, 2008; Adler *et al.*, 2014] and *Epithemia* [Nakayama *et al.*, 2011; Nakayama and Archibald, 2012; Nakayama and Inagaki, 2014; Nakayama *et al.*, 2014]. They lack the genes for PSII, phycobilisomes and Chl synthesis, and the ability for CO_2 assimilation has also been lost. The *Epitemia* symbiont is the most reduced one, and lacks, in addition to the above, genes for PSI [Nakayama *et al.*, 2014]; on the other hand, there are also cases of less reduced cyanobacterial symbionts inside some diatoms [Carpenter and Janson, 2000].

Another somewhat similar case is the symbiosis between UCYN-A cyanobacteria lacking PSII and prymnesiophytes (haptophytes) [Thompson *et al.*, 2012; Hagino *et al.*, 2013; Thompson *et al.*, 2014; Krupke *et al.*, 2015; Cabello *et al.*, 2016]. Here also, the cyanobacterial symbiont is very reduced. Although it is unable to assimilate CO_2, it does assimilate N_2. In another symbiosis, described by Hagino *et al.* [2013], the prokaryote is inside the cell of the eukaryote. However, Thomson *et al.* [2012] and Krupke *et al.* [2015] have described other examples, where prokaryotes are firmly and specifically attached to the outside of eukaryotes.

8.2   Adapting Photosynthesis that has Evolved Underwater for Life on Land

The photosynthetic system, originally evolved in an aquatic environment, must have been exposed to a number of challenges when the organisms containing them were colonizing land. The most important ones are:

1. The ultimate reductant (electron donor) in oxygenic photosynthesis is water; thus, water must still be available. Apart from that, water must be available to provide an environment for various enzymes and their reactions.
2. In some terrestrial environments, the episodic loss of water cannot be avoided, and under such circumstances the system must acquire the ability to tolerate temporary desiccation.
3. Temperature fluctuations are greater in an aerial environment than in an aquatic one, and the system must be resilient to such fluctuations.
4. The light environment may be much more variable, with occasional shading by nearby objects. Thus, the system should be able to cope with changes in light, not only the daily light rhythm.
5. The spectrum of ambient light is different in terrestrial than in aquatic environment. Further, much more red and far-red light is available on the land than under water. When land plants, or parts of complex canopies, shade one another, the effect of far-red light may completely dominate.

8.2.1 *Grana*

One remarkable structural feature with which the photosynthetic system of plants stands out, in comparison to that of aquatic organisms, is the presence of appressed thylakoids and grana. With the exception of bundle-sheath chloroplasts of C4 plants of the NADP-malic enzyme type, grana are almost universally present in chloroplasts of embryophytes (bryophytes, pteridophytes, and spermatophytes), which dominate the land vegetation. Can we find a connection between the occurrence of grana and the differences between the aquatic and aerial/terrestrial environments listed above?

This question has occupied the minds of a number of researchers [Andersson and Anderson, 1980; Trissl and Wilhelm, 1993; Mullineaux, 2005; Anderson *et al.*, 2008; Anderson, 2012; Nevo *et al.*, 2012], and several of them refer to the differences between PSI and PSII, and the cooperation between the two photosystems. They point to the need for controlling the "drain" of energy from the intrinsically somewhat slower PSII to PSI, and thus a need for the separation of the two photosystems: PSII mostly in the grana, and PSI mostly in "stroma thylakoids" ("intergranal thylakoids") (Fig. 2.4). PSI has a somewhat deeper, irreversible, reaction center trap, and longer-wavelength antenna (providing the opportunity for spill-over of excitation energy form PSII to PSI). In the aquatic environment, long-wavelength radiation is absorbed by water, and the risk for over-excitation of PSI is therefore unlikely. It has also been argued that grana provide space for the larger PSII antenna systems [Pribil *et al.*, 2014].

Apart from their lack (or lesser development) of grana, the bundle-sheath chloroplasts of NADP-malic enzyme type C4 plants (Fig. 4.6) are notable because they lack or have weakly developed O_2-evolution system (see [Romanowska *et al.*, 2006] and references therein). This avoids competition by O_2 with CO_2 for RuBisCO. In some of these plants, regular PSII function is absent, and the energy captured by PSII is channeled on to PSI [Pfundel and Pfeffer, 1997]. This type of C4 plants has no need for non-cyclic electron transfer, and thus no need for PSII activity, since the reduction of NADP is achieved by oxidation of malate by malic enzyme (Fig. 4.6). So obviously these chloroplasts do not have any need for grana.

Charophytes are the closest algal relatives of terrestrial plants. Most charophytes are restricted to either marine or freshwater environments, but there are some, among the contemporary species, which had begun the journey onto land. Charophytes of the genus *Entransia* are found in humid habitats such as *Sphagnum* bogs; *Hormidiella* grows in moist soil or temporary ditches of rainwater, while *Klebsormidium* and *Interfilum* species are found in places that often dry out [Kitzing and Karsten, 2015]. When exposed to UV radiation, these plants accumulate mycosporine-like amino acids, as "sunscreens", a trait shared with many algae. Higher land plants, but not bryophytes, use flavonoids, for this purpose [Rozema *et al.*, 2002]. Most of the cell-wall components typical of embryophytes are already present in charophytes [Mikkelsen *et al.*, 2014]. The ability of charophytes to deposit callose (plant polysaccharides) seems to correlate with their ability to survive drought [Herburger and Holzinger, 2015]. Not surprisingly, charophytes exhibit some traits typical for algae, but also others, which are usually considered typical for land plants. Uniquely, among algae, charophyte chloroplasts have grana [Gunning and Schwartz, 1999; Vouilloud *et al.*, 2015], which otherwise occur only in the land plants from the liverworts [Sun, 1963] to the seed plants. Typical grana are absent from other algae, although somewhat grana-like structures are found in, *e.g.*, *Chlamydomonas* chloroplasts [Engel *et al.*, 2015].

8.2.2 Hormones

Synthesis of plant hormones, such as auxins, ethene, abscisic acid, cytokinin, and giberellic acid, has been shown to take place not only in land plants, but in a number of other organisms: cyanobacteria, multicellular brown algae, multicellular red algae, and green algae [Lu and Xu, 2015]. This is not the same as saying that these organisms can use these compounds for signaling. Most of the well-known plant hormones acquired a signaling function late in plant evolution [Wang *et al.*, 2015]. A number of other components, not discussed here, are also required for signal sensing and transduction to function.

However, the charophyte *Klebsormidium* had an auxin signaling system, which is typical of land plants [Wang *et al.*, 2015]. In a member of this genus, Hori *et al.* [2014] discovered genes for the synthesis of well-known plant hormones abscisic acid and jasmonic acid. We shall in the following section concentrate on abscisic acid, as this is involved in the adaptation of plants to the scarcity of water, which may occur in the terrestrial environment. In cyanobacteria, algae, and lichens, drought or salt stress induces the synthesis of abscisic acid, although the signaling pathway for this hormone has not yet been established (see *e.g.*, [Hartung, 2010]). In land plants its synthesis is induced by various stresses, but in particular by drought.

The effect that has given abscisic acid its name is the abscission of leaves and other organs. Leaf abscission helps a plant exposed to drought to conserve water. Abscisic acid also prevents seeds from germinating until water supply is sufficient for continued growth. A third important way in which abscisic acid is involved in helping non-aquatic plants to conserve their inner water is in the regulation of stomata (see below).

We end this section by mentioning another type of signaling molecules in connection with adaptation to land life: they are the strigolactones, mediating cooperation of plants with fungi, which has been absolutely necessary for the establishment of plants on land. Strigolactone signaling can be traced back to the charophytes [Delaux *et al.*, 2012], which were therefore also in this respect pre-adapted for a conquest of the land surface [Delaux *et al.*, 2015].

8.2.3 *Stomata*

As formulated by Raschke [1976] and recently repeated by Leung *et al.* [2016]:

> Land plants are in a dilemma throughout their lives: assimilation of CO_2 from the atmosphere requires intensive gas exchange; the prevention of excessive water loss demands that gas exchange be kept low.

Land plants have addressed this dilemma by covering themselves with a cuticle that diminishes evaporation; it is penetrated by stomatal pores that are regulated by a very complex signaling system.

Cook and Graham [1998] found structural similarities between the surface layer of the charophycean *Coleochaete orbicularis* and the initial pro-cuticle described in a model of cuticle development in higher plants [Jeffree, 1996]. However, no mechanistic information, in terms of chemistry, is available. Further, we note that a bryophyte has been shown to have a cuticle with a chemical composition similar to that of a seed plant [Busta *et al.*, 2016].

Stomatal aperture is governed by the turgidity of the surrounding guard cells, and this, in turn, is regulated in a number of ways according to light conditions, internal CO_2 concentration, and water status, as well as with a component of circadian rhythmicity. With the exception of CAM plants, it is advantageous for plants to have stomata closed when light is not sufficient for photosynthesis, or when they are exposed to risk of drying out. Various aspects of this regulation process have been described by Kim *et al.* [2010] and Murata *et al.* [2015].

Stomata of the modern land plants are very similar to those of the seed plants found in *ca.* 400 Ma old fossils of the early embryophytes [Edwards *et al.*, 1998; Hanson and Rice, 2014]. Similar stomata are also found in extant mosses [Merced and Renzaglia, 2014], and these respond to abscisic acid, just as stomata of seed plants do [Chater *et al.*, 2013]. The blue-light-specific rapid opening response associated with K^+ accumulation is found not only in spermatophytes, but also in non-spermatophyte vascular plants, such as *Equisetum hyemale* and *Psilotum nudum* [Doi *et al.*, 2015].

In most spermatophytes, guard cells are the only epidermal cells that contain chloroplasts, and their photosynthesis is necessary for providing the turgor necessary for keeping the stomata open [Azoulay-Shemer *et al.*, 2015]. Chloroplasts are also the site for the blue-light response [Horrer *et al.*, 2016].

Plant roots sense the water availability in the soil and, if it is insufficient, they synthesize abscisic acid and send it as a closing signal to the stomata in the leaves. But guard cell chloroplasts also make abscisic acid for inducing closure of the stomata; further, high CO_2-concentration causes closing of stomata *via* the activity of carbonic anhydrase [Hu *et al.*, 2015; Engineer *et al.*, 2016] as well as of abscisic acid. However, the regulation of the opening and closing of stomata by abscisic acid appears to be absent in the ferns and in the lycophytes [McAdam and Brodribb, 2012].

8.2.4   *Reaching for light*

8.2.4.1   *Evolution of trees*

When soil accumulated and conditions on the land improved, plants started to compete with one another for light. Many outmaneuvered the competitors by growing taller [Li *et al.*, 2015]. Thus, the first trees evolved. The earliest known trees, from the genus *Wattieza*, appeared in the Late Devonian, *ca.* 385 Ma ago.

8.2.4.2   *Cheating: Lianas (Climbers) and Epiphytes*

Probably soon after the invention of arborescent growth, the "cheaters" appeared. ("Cheating" is defined as a trait that is beneficial to a cheat, but costly to a "cooperator" in terms of inclusive fitness.) For a detailed discussion and understanding, see Ghoul *et al.* [2013]. "Cheaters" include 537 different species of climbing plants, known from the Carboniferous [Burnham, 2009]. Further, the climbing pteridosperms (seed ferns) from the Carboniferous and early Permian are "cheaters" and they have been described, in detail, by Krings *et al.* [2003]. Also included in the group are epiphytes, known already from the Carboniferous era [Psenicka and Oplustil, 2013].

8.3   The Role of Fungi

Although not always recognized, fungi have been important actors in the conquest of land by plants. By the formation of lichens they first helped cyanobacteria and algae to take the first step to go on to the land. Later came the plants proper, always in different kinds of association with fungi. In the contemporary flora, 90% of the plant species have mycorrhizal associations, and many have endophytic fungi in their shoots (*e.g.*, [Hardoim *et al.*, 2015, Dastias *et al.*, 2017]). Today lichens are widely employed for ecosystem monitoring [Shukla *et al.*, 2014].

8.4 Making More Photosynthesis... More Biomass, More Bioenergy, New Chemicals and Hydrogen

8.4.1 *General: Classical breeding and genetic engineering*

A limiting factor for the productivity of C3 plants under many conditions is the low affinity of RuBisCO for CO_2 and the competing affinity for O_2. Attempts to improve this situation can be classified into three categories: Conversion of C3 crops to C4 crops by genetic engineering; installing CO_2-concentrating mechanisms used by cyanobacteria or algae, and improving the properties of RuBisCO.

8.4.2 *Conversion from C3 to C4 metabolism by genetic engineering*

At first sight attempting to introduce an efficient C4 pathway into C3 plants might appear an impossible task, as it requires not only the introduction of a new metabolic cycle, but also several other changes. Considering that natural evolution has resulted in such conversion many dozens of times makes it appear a little more feasible [Covshoff and Hibberd, 2012]. Since there exist organisms with single-cell C4 metabolism, one might think that it would be easier to engineer such a system, as it would not require any new transport mechanisms between cells. However, despite attempts during many years, this strategy has not resulted in much progress (e.g., see [Taniguchi *et al.*, 2008]).

A current focus is to engineer NADP-malic enzyme type C4 pathway into rice, a crop plant of enormous importance for a large part of Earth's population [Karki *et al.*, 2013]. A special international consortium, the "C4 rice consortium" with about 20 universities and research institutes including the International Rice Research Institute in the Philippines, and also several other research groups are working toward this goal. Contrary to what one might expect, the main "culprit" is the decision not to introduce enzymes for the C4 cycle. This is because all of the genes necessary for this are already present in C3 plants, even if there is reason to change some of the enzymes to be better suited for C4 pathway.

Researchers have attempted to pinpoint the transcriptome characteristics of C3 and C4 mesophyll cells, and the bundle-sheath cells, by comparing the gene expression in closely related C3 and C4 plants, *e.g.*, in the genera *Flaveria* [Gowik *et al.*, 2011] and *Cleome* [Marshall *et al.*, 2007], in plants that alternate between C3- and C4-pathways depending on conditions, and also in organs of the same plant that differ in their metabolic systems. Comparison of gene expressions in bundle-sheath and mesophyll cells [John *et al.*, 2014] has resulted in knowledge about important transcription factors. Young leaf cells of *Cleome gynandra* perform C3 photosynthesis, but switch to C4, and study of changes in gene expression and transcription factors during leaf development, in this system, has provided valuable information [Aubry *et al.*, 2014].

A couple of examples of increased photosynthesis and biomass by genetic engineering are given here: (*i*) Overexpression of PEP carboxylase, a C-4 enzyme from maize in a model plant *Arabidopsis thaliana* led to increased electron transfer rate; lowered loss by energy as heat, higher performance and increased dry weight [Kandoi *et al.*, 2016]. Attempts are being made to transfer this technology to rice. (*ii*) Soda *et al.* [2018] showed that in a transgenic rice, rice intermediate filament stabilized photosynthetic machinery and yield under salinity and heat stress.

The following tasks are the targets for some of the future research:

- Increase the density of vascular bundles [Feldman *et al.*, 2014]; a reason for this is that bundle-sheath cells must be close to other mesophyll cells for efficient exchange of metabolites;
- Limit the expression of RuBisCO genes to cells close to the vascular bundles;
- Establish photorespiratory bypasses, and much more.

8.4.3 *Cyanobacterial genes in higher plants*

Earlier attempts to improve plant properties using cyanobacterial genes have been reviewed by Park *et al.* [2009], and we mention here only some recent work. Cyanobacteria have transporters for CO_2 and HCO_3^-, and a carboxysome structure that maintains a high CO_2 concentration at the RuBisCO. Lin *et al.* [2014] took the first steps for constructing a cyanobacterial carboxysome in *Nicotiana* chloroplasts. However, in a theoretical analysis, McGrath and Long [2014] arrived at the conclusion that providing a carboxysome structure alone would decrease the maximum photosynthesis rate, and that the best first step should be to introduce transporters. Introduction of just one transporter could result in a 9% increase, and that all of them a 60% increase. Uehara *et al.* [2016] have already installed a cyanobacterial HCO_3^- transporter in *Arabidopsis*, but the physiological effects have not yet been evaluated.

8.4.4 *Improvement of RuBisCO*

Four properties of RuBisCO are seen as targets for improvement: (*1*) the low affinity for CO_2; (*2*) affinity for O_2; (*3*) slow turnover; and, (*4*) inhibition by xylulose-1,5-bisphosphate (XuBP). These properties are not independent of each other. A higher turnover rate appears to be associated with lower affinity for CO_2 and less discrimination against O_2 in favor of CO_2 [Savir *et al.*, 2010]. Therefore, the way to go forward seems to select for a fast RuBisCO and solve the problems (*1*) and (*2*) by introduction of C4 pathways or other CO_2 concentration mechanisms.

An important discovery has been the competitive inhibition of RuBisCO by XuBP and the role of RuBisCO activase in removing this inhibitor [Portis, 2003; Parry *et al.*, 2008]. XuBP is formed from RuBP on the substrate-binding site of RuBisCO from RuBP at the rate of 0.25% of that of the carboxylation of RuBP [Pierce, 1989; Robinson and Portis, 1989]. XuBP and RuBisCO, separated by the activase, can recombine, so that a more persistent RuBisCO activation could be achieved by breaking down XuBP. Bracher *et al.* [2015] found that this can be achieved by incorporating a phosphatase from *Rhodobacter sphaeroides*, an anoxygenic photosynthetic bacteria. This dephosphorylates XuBP to xylulose-5-phosphate, which through the Calvin-Benson cycle enzymes can be recycled to RuBP. For current achievements in increasing carbon fixation (carbohydrate production) either through the Calvin-Benson cycle, or by the introduction of the algal pyrenoid into higher plants, see Sharwood [2017a,b] and references therein.

8.4.5   *Photorespiratory bypass*

In C3 plants, up to 30% of newly assimilated carbon is lost by photorespiration, and some nitrogen may also be lost by release of ammonia. Inhibition of enzymes involved in photorespiration is not a feasible way to decrease this loss [Xin *et al.*, 2015]. A better way is to modify photorespiratory metabolism. One possibility is to convert glycolate to glycerate [Kebeish *et al.*, 2007], which abolishes ammonia release and increases net photosynthesis. Dalal *et al.* [2015], based on this concept, constructed a modified version of the biofuel crop *Camelina sativa* with increased yield. Another way, so far not successfully realized, to prevent ammonia release is to convert glyoxylate formed in the peroxysome to hydroxypyruvate [Carvalho *et al.*, 2011]. Maier *et al.* [2012] employed a third version of photorespiratory bypass (Fig. 8.3). By introducing a glycolate oxidase into the chloroplast (and a catalase to prevent damage by the hydroperoxide formed) in combination with a malic synthetase, they could reduce NADP and recover CO_2 inside the chloroplast. *Arabidopsis* plants engineered in this way showed increased biomass production. For further information on advances and future perspectives in photorespiratory engineering and manipulations, see Betti *et al.* [2016].

8.4.6   *Photosynthetic H_2 production*

Conversion of the unlimited supply of sunlight into an ecologically "clean" fuel, such as molecular hydrogen (H_2) is one of the most promising research directions towards solving the increasing energy demand. How can photosynthetic organisms help us in this regard?

The first indication about the existence of an enzyme *hydrogenase*, which converts protons and electrons into molecular hydrogen (H_2) in a photosynthetic system, was obtained by Gaffron and Rubin [1942] more than 75 years ago. They observed light-induced production of H_2 by a green alga, *Scenedesmus obliquus*, after it was dark-adapted under CO_2-free conditions. More than 40 years later, Roessler and Lien [1984] found that reduced ferredoxin (Fd) is the direct electron donor to hydrogenase in green algae (Fig. 8.4). Today, we know that not only green algae, but also a large number of cyanobacteria are capable of photosynthetic H_2

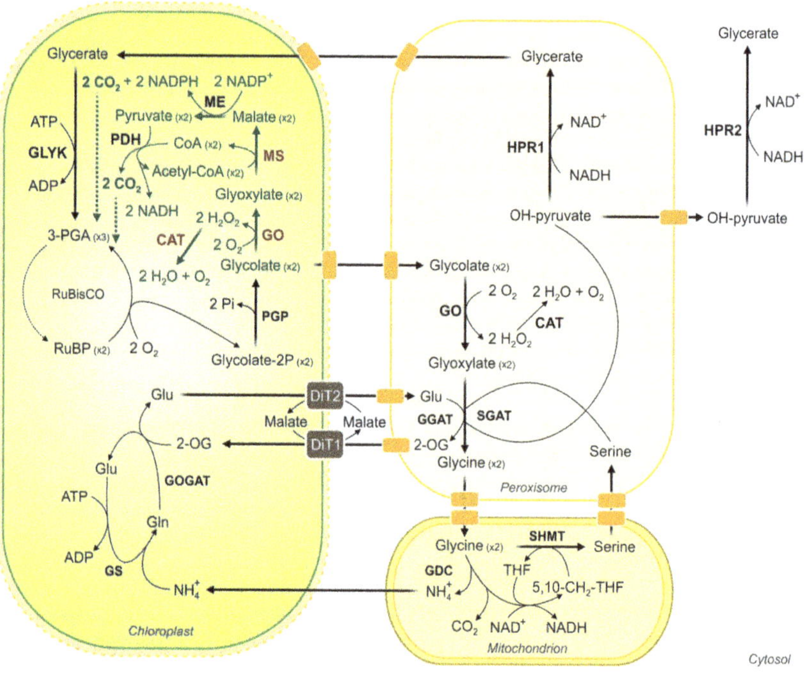

Fig. 8.3. Metabolism in *Arabidopsis thaliana* engineered with the Maier *et al.* [2012] photorespiratory bypass in the chloroplast shown in green, with the acronyms of the newly introduced enzymes in dark red (GO, glycolate oxidase; CAT, catalase; MS, malic synthetase). Other enzymes in the cycle are NADP-malic enzyme (ME) and pyruvate dehydrogenase (PDH).

production (for reviews, see [Esper *et al.*, 2006; Ghirardi *et al.*, 2007; Lubitz *et al.*, 2008; Ghirardi *et al.*, 2009; Hemschemeier *et al.*, 2009; Zannoni and De Philippis, 2014; Rögner, 2015]). Thus, using light as a driving force to extract electron from water, these organisms generate strong reductants, such as reduced Fd (in case of algae) and NADPH (in case of cyanobacteria), which can be utilized as substrates for H_2 production by hydrogenases (Fig. 8.4). Moreover, in addition to hydrogenases, nitrogenases can also catalyze generation of H_2 in cyanobacteria and in anoxygenic purple bacteria. However, since nitrogenases require ATP for such a process, the production of H_2 by this enzyme is energetically more "expensive" than by hydrogenases (see Section 7.4 and Fig. 7.5 in Chapter 7, also see reviews, by Ghirardi *et al.* [2009] and by Bothe *et al.* [2010]). But even in case of

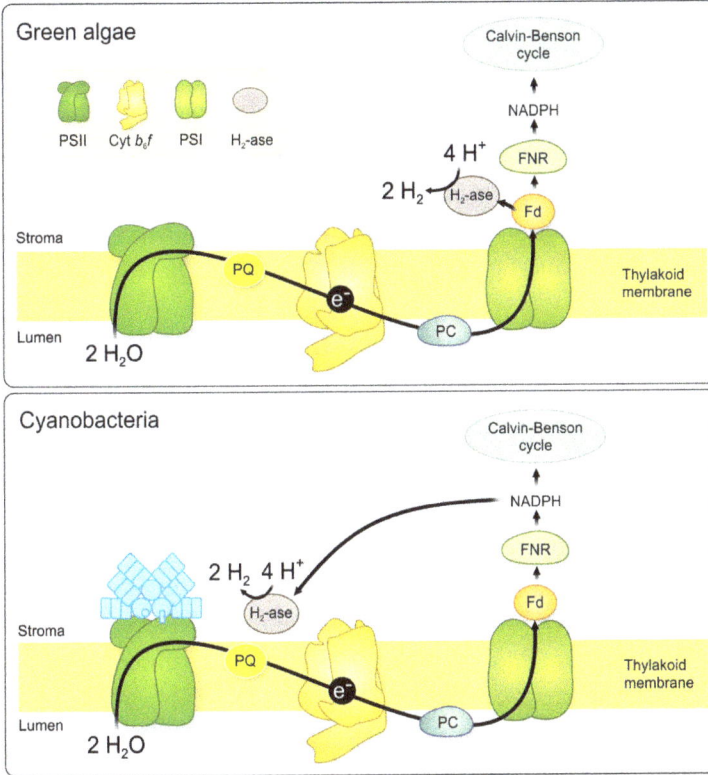

Fig. 8.4. Simplified schematic representation of photosynthetic H_2 production pathways catalyzed by hydrogenases in green algae (top) and cyanobacteria (bottom). Abbreviations: PSII, photosystem II; PQ, plastoquinone; Cyt b_6f, cytochrome b_6f complex, PC, plastocyanin; PSI, photosystem I; Fd, ferredoxin; FNR, ferredoxin–NADP reductase; H_2-ase, hydrogenase.

hydrogenases, one needs to overcome several biochemical/metabolic barriers [Tamagnini *et al.*, 2007; Oey *et al.*, 2016] and even some technical barriers [Sharma and Arya, 2017] for sustained and economically feasible H_2 production. The most important among of them are the following:

- The irreversible inhibition of most hydrogenases by O_2;
- The utilization of photosynthetically produced H_2 by bidirectional hydrogenases;
- The competition of H_2 production with the Calvin–Benson cycle for the reducing equivalents (*e.g.*, NADPH).

151

Due to these restrictions the H_2 production in wild-type photosynthetic organisms takes place only under special conditions, resulting in very low yields. To overcome these limitations, and thus, to improve the rates of H_2 evolution, biological (genetic) and technical (photobioreactor) engineering, as well as other techniques are being employed by many research groups [Lubitz *et al.*, 2008; Rögner, 2015; Martin and Frymier, 2017; Seibert and Tor-zillo, 2018]. Nevertheless, photobiological H_2 production has not yet reached economic viability thus far [Blankenship *et al.*, 2011; Oey *et al.*, 2016].

8.4.7 *Biofuels and other chemicals*

Biofuel production from biomass (crops or algae) is thought to have a potential to be globally used as renewable energy source in the future. In addition, gradually reducing usage of limited fossil fuels, biofuels, is expected to reduce the amounts of greenhouse gases in the Earth's atmosphere. Moreover, biomass and biofuel can also be used for the production of bio-plastics [Balaji *et al.*, 2013] or other chemicals [Jeon *et al.*, 2017]. For details on current challenges in biofuel research, its production, and for its future perspectives, see, Hannon *et al.* [2010]; Sayre [2010]; Kumar *et al.* [2018].

8.4.8 *Other improvements*

We end this Chapter by listing several approaches expected to improve ("tweak") photosynthesis for the benefit of all of us. These approaches include:

- Decreasing antenna size of the photosystems in order to increase photosynthetic productivity by having a larger proportion of the absorbed photons being converted to biomass [Ort *et al.*, 2015; Kirst *et al.*, 2017];
- Engineering organisms (or employing wild-type organisms) in order to capture and utilize extreme far-red light [Thapper *et al.*, 2009; Blankenship *et al.*, 2011; Chen and Blankenship, 2011];
- Improving protection and recovery mechanisms against photodamage by excess light [Kromdijk *et al.*, 2016].

For further information, we refer the readers to an excellent editorial by Bailey-Serres *et al.* [2018] and references therein. Attempts, recent achivements, and future perspectives in developing artificial photosynthetic systems will be discussed in Chapter 9, the last chapter of this book.

References

Adler, S., Trapp, E. M., Dede, C., Maier, U. G. and Zauner, S. (2014). Rhopalodia gibba: The First Steps in the Birth of a Novel Organelle? *In* Löffelhardt, W., ed, *Endosymbiosis*. (Springer, Vienna), pp. 167–179.

Anderson, J. M. (2012). Lateral heterogeneity of plant thylakoid protein complexes: Early reminiscences, *Philos. Trans. R. Soc. Lond., B*, 367, pp. 3384–3388.

Anderson, J. M., Chow, W. S. and De Las Rivas, J. (2008). Dynamic flexibility in the structure and function of photosystem II in higher plant thylakoid membranes: The grana enigma, *Photosynth. Res.*, 98, pp. 575–587.

Andersson, B. and Anderson, J. M. (1980). Lateral heterogeneity in the distribution of chlorophyll-protein complexes of the thylakoid membranes of spinach chloroplasts, *Biochim. Biophys. Acta*, 593, pp. 427–440.

Aubry, S., Kelly, S., Kümpers, B. M. C., Smith-Unna, R. D. and Hibberd, J. M. (2014). Deep evolutionary comparison of gene expression identifies parallel recruitment of trans-factors in two independent origins of C4 photosynthesis, *PLos Genet.*, 10, e1004365.

Azoulay-Shemer, T., Palomares, A., Bagheri, A., Israelsson-Nordstrom, M., Engineer, C. B., Bargmann, B. O. R., Stephan, A. B. and Schroeder, J. I. (2015). Guard cell photosynthesis is critical for stomatal turgor production, yet does not directly mediate CO_2- and ABA-induced stomatal closing, *Plant J.*, 83, pp. 567–581.

Bailey-Serres, J., Pierik, R., Ruban, A. and Wingler, A. (2018). The dynamic plant: Capture, transformation, and management of energy, *Plant Physiol.*, 176, pp. 961–966.

Balaji, S., Gopi, K. and Muthuvelan, B. (2013). A review on production of poly β hydroxybutyrates from cyanobacteria for the production of bio plastics, *Algal Res.*, 2, pp. 278–285.

Ball, S. G., Bhattacharya, D. and Weber, A. P. M. (2016). Pathogen to powerhouse, *Science*, 351, pp. 659–660.

Ball, S. G., Colleoni, C., Kadouche, D., Ducatez, M., Arias, M.-C. and Tirtiaux, C. (2015). Toward an understanding of the function of Chlamydiales in plastid endosymbiosis, *Biochim. Biophys. Acta*, 1847, pp. 495–504.

Ball, S. G., Subtil, A., Bhattacharya, D., Moustafa, A., Weber, A. P. M., Gehre, L., Colleoni, C., Arias, M.-C., Cenci, U. and Dauvillée, D. (2013). Metabolic effectors secreted by bacterial pathogens: Essential facilitators of plastid endosymbiosis?, *Plant Cell*, 25, pp. 7–21.

Bastias, D. A., Martínez-Ghersa, M. A., Ballaré, C. L. and Gundel, P. E. (2017). Epichloë fungal endophytes and plant defenses: Not just alkaloids, *Trends Plant Sci.*, 22, pp. 939–948.

Betti, M., Bauwe, H., Busch, F. A., Fernie, A. R., Keech, O., Levey, M., Ort, D. R., Parry, M. A. J., Sage, R., Timm, S., Walker, B., Weber, A. P. M. (2016). Manipulating photorespiration to increase plant productivity: Recent advances and perspectives for crop improvement, *J. Exp. Bot.*, 67, pp. 2977–2988.

Bennett, M. S. and Triemer, R. E. (2015). Chloroplast genome evolution in the Euglenaceae, *J. Eukaryot. Microbiol.*, 62, pp. 773–785.

Blank, C. E. (2013). Origin and early evolution of photosynthetic eukaryotes in freshwater environments. Reinterpreting proterozoic paleobiology and biochemical processes in light of trait evolution, *J. Phycol.*, 49, pp. 1040–1055.

Blankenship, R. E., Tiede, D. M., Barber, J., Brudvig, G. W., Fleming, G. R., Ghirardi, M. L., Gunner, M. R., Junge, W., Kramer, D. M., Melis, A., Moore, T. A., Moser, C. C., Nocera, D. G., Nozik, A. J., Ort, D. R., Parson, W. W., Prince, R. C. and Sayre, R. T. (2011). Comparing photosynthetic and photovoltaic efficiencies and recognizing the potential for improvement, *Science*, 332, pp. 805–809.

Bothe, H., Schmitz, O., Yates, M. G. and Newton, W. E. (2010). Nitrogen fixation and hydrogen metabolism in cyanobacteria, *Microbiol. Mol. Biol. Rev.*, 74, pp. 529–551.

Bracher, A., Sharma, A., Starling-Windhof, A., Hartl, F. U. and Hayer-Hartl, M. (2015). Degradation of potent Rubisco inhibitor by selective sugar phosphatase, *Nature Plants*, 1, 14002.

Burnham, B. J. (2009). An overview of the fossil record of climbers: Bejucos, Sogas, Trepadoras, Lianans, Cipós, and Vines, *Rev. Bras. Palaeontol.*, 12, pp. 149–160.

Busta, L., Budke, J. M. and Jetter, R. (2016). Identification of β-hydroxy fatty acid esters and primary, secondary-alkanediol esters in cuticular waxes of the moss *Funaria hygrometrica*, *Phytochemistry*, 121, pp. 38–49.

Cabello, A. M., Cornejo-Castillo, F. M., Raho, N., Blasco, D., Vidal, M., Audic, S., de Vargas, C., Latasa, M., Acinas, S. G. and Massana, R. (2016). Global distribution and vertical patterns of a prymnesiophyte–cyanobacteria obligate symbiosis, *ISME J.*, 10, pp. 693–706.

154

Carpenter, E. J. and Janson, S. (2000). Intracellular cyanobacterial symbionts in the marine diatom *Climacodium frauenfeldianum* (Bacillariophyceae), *J. Phycol.*, 36, pp. 540–544.

Carvalho, J. F. C., Madgwick, P. J., Powers, S. J., Keys, A. J., Lea, P. J. and Parry, M. A. J. (2011). An engineered pathway for glyoxylate metabolism in tobacco plants aimed to avoid the release of ammonia in photorespiration, *BMC Biotechnol.*, 11, 111.

Chater, C., Gray, J. E. and Beerling, D. J. (2013). Early evolutionary acquisition of stomatal control and development gene signalling networks, *Curr. Opin. Plant Biol.*, 16, pp. 638–646.

Chen, M. and Blankenship, R. (2011). Expanding the solar spectrum used by photosynthesis, *Trends Plant Sci.*, 16, pp. 427–431.

Cook, M. E. and Graham, L. E. (1998). Structural similarities between surface layers of selected Charophycean algae and Bryophytes and the cuticles of vascular plants, *Int. J. Plant Sci.*, 159, pp. 780–787.

Covshoff, S. and Hibberd, J. M. (2012). Integrating C4 photosynthesis into C3 crops to increase yield potential, *Curr. Opin. Biotechnol.*, 23, pp. 209–214.

Dalal, J., Lopez, H., Vasani, N. B., Hu, Z., Swift, J. E., Yalamanchili, R., Dvora, M., Lin, X., Xie, D., Qu, R. and Sederoff, H. W. (2015). A photorespiratory bypass increases plant growth and seed yield in biofuel crop *Camelina sativa*, *Biotechnol. Biofuels*, 8, 175.

Delaux, P.-M., Radhakrishnan, G. V., Jayaraman, D., Cheema, J., Malbreil, M., Volkening, J. D., Sekimoto, H., Nishiyama, T., Melkonian, M., Pokorny, L., Rothfels, C. J., Sederoff, H. W., Stevenson, D. W., Surek, B., Zhang, Y. R., Sussman, M. R., Dunand, C., Morris, R. J., Roux, C., Wong, G. K.-S., Oldroyd, G. E. D. and Ané, J.-M. (2015). Algal ancestor of land plants was preadapted for symbiosis, *Proc. Natl. Acad. Sci. U.S.A.*, 112, pp. 13390–13395.

Delaux, P.-M., Xie, X., Timme, R. E., Puech-Pages, V., Dunand, C., Lecompte, E., Delwiche, C. F., Yoncyama, K., Bécard, G. and Séjalon-Delmas, N. (2012). Origin of strigolactones in the green lineage, *New Phytol.*, 195, pp. 857–871.

Doi, M., Kitagawa, Y. and Shimazaki, K.-I. (2015). Stomatal blue light response is present in early vascular plants, *Plant Physiol.*, 169, pp. 1205–1213.

Domman, D., Horn, M., Embley, T. M. and Williams, T. A. (2015). Plastid establishment did not require a chlamydial partner, *Nature Commun.*, 6, 6421.

Edwards, D., Kerp, H. and Hass, H. (1998). Stomata in early land plants: An anatomical and ecophysiological approach, *J. Exp. Bot.*, 49, pp. 255–278.

Engel, B. D., Schaffer, M., Kuhn Cuellar, L., Villa, E., Plitzko, J. M. and Baumeister, W. (2015). Native architecture of the *Chlamydomonas* chloroplast revealed by in situ cryo-electron tomography, *eLife*, 4, e04889.

Engineer, C. B., Hashimoto-Sugimoto, M., Negi, J., Israelsson-Nordström, M., Azoulay-Shemer, T., Rappel, W.-J., Iba, K. and Schroeder, J. I. (2016). CO_2 sensing and CO_2 regulation of stomatal conductance: Advances and open questions, *Trends Plant. Sci.*, 21, pp. 16–30.

Esper, B., Badura, A. and Rögner, M. (2006). Photosynthesis as a power supply for (bio-)hydrogen production, *Trends Plant. Sci.*, 11, pp. 543–549.

Feldman, A. B., Murchie, E. H., Leung, H., Baraoidan, M., Coe, R., Yu, S.-M, Lo, S.-F. and Quick, W. P. (2014). Increasing leaf vein density by mutagenesis: Laying the foundations for C4 Rice, *PLoS ONE*, 9, e94947.

Gaffron, H. and Rubin, J. (1942). Fermentative and photochemical production of hydrogen in algae, *J. Gen. Physiol.*, 26, pp. 219–240.

Ghoul, M., Griffin, A. S. and West, S. A. (2013). Toward an evolutionary definition of cheating, *Evolution*, 68, *pp. 318–331.*

Ghirardi, M. L., Posewitz, M. C., Maness, P. C., Dubini, A., Yu, J. and Seibert, M. (2007). Hydrogenases and hydrogen photoproduction in oxygenic photosynthetic organisms, *Annu. Rev. Plant Biol.*, 58, pp. 71–91.

Ghirardi, M. L., Dubini, A., Yu, J. and Maness, P.-C. (2009). Photobiological hydrogen-producing systems, *Chem. Soc. Rev.*, 38, pp. 52–61.

Gowik, U., Bräutigam, A., Weber, K. L., Weber, A. P. M. and Westhoff, P. (2011). Evolution of C_4 photosynthesis in the genus *Flaveria*: How many and which genes does it take to make C_4?, *Plant Cell*, 23, pp. 2087–2105.

Gunning, B. E. S. and Schwartz, O. M. (1999). Confocal microscopy of thylakoid autofluorescence in relation to origin of grana and phylogeny in the green algae, *Funct. Plant Biol.*, 26, pp. 695–708.

Hagino, K., Onuma, R., Kawachi, M. and Horiguchi, T. (2013). Discovery of an endosymbiotic nitrogen-fixing cyanobacterium UCYN-A in *Braarudosphaera bigelowii* (Prymnesiophyceae), *PLoS ONE*, 8, e81749.

Hannon, M., Gimpel, J., Tran, M., Rasala, B. and Mayfield, S. P. (2010). Biofuels from algae: Challenges and potential, *Biofuels*, 1, pp. 763–784.

Hanson, D. T. and Rice, S. K., eds (2014). *Photosynthesis in Bryophytes and Early Land Plants*, (Springer Netherlands, Dordrecht).

Händeler, K., Grzymbowski, Y. P., Krug, P. J. and Wägele, H. (2009). Functional chloroplasts in metazoan cells - a unique evolutionary strategy in animal life, *Front. Zool.*, 6, 28.

Hardoim, P. R., van Overbeek, L. S., Berg, G., Pirttilä, A. M., Compant, S., Campisano, A., Döring, M. and Sessitsch, A. (2015). The hidden world within plants: Ecological and evolutionary considerations for defining functioning of microbial endophytes, *Microbiol. Molec. Biol. Revs*, 79, pp. 293–320.

Hartung, W. (2010). The evolution of abscisic acid (ABA) and ABA function in lower plants, fungi and lichen, *Funct. Plant Biol.*, 37, pp. 806–812.

Hemschemeier, A., Melis, A. and Happe, T. (2009). Analytical approaches to photobiological hydrogen production in unicellular green algae, *Photosynth. Res.*, 102, pp. 523–540.

Herburger, K. and Holzinger (2015). Localization and quantification of callose in the streptophyte green algae *Zygnema* and *Klebsormidium*: Correlation with desiccation tolerance, *Plant Cell Physiol.*, 56, pp. 2259–2270.

Hess, W. R., Partensky, F., van der Staay, G. W., Garcia-Fernandez, J. M., Borner, T. and Vaulot, D. (1996). Coexistence of phycoerythrin and a chlorophyll a/b antenna in a marine prokaryote, *Proc. Natl. Acad. Sci. U.S.A.*, 93, pp. 11126–11130.

Hori, K., Maruyama, F., Fujisawa, T., Togashi, T., Yamamoto, N., Seo, M., Sato, S., Yamada, T., Mori, H., Tajima, N., Moriyama, T., Ikeuchi, M., Watanabe, M., Wada, H., Kobayashi, K., Saito, M., Masuda, T., Sasaki-Sekimoto, Y., Mashiguchi, K., Awai, K., Shimojima, M., Masuda, S., Iwai, M., Nobusawa, T., Narise, T., Kondo, S., Saito, H., Sato, R., Murakawa, M., Ihara, Y., Oshima-Yamada, Y., Ohtaka, K., Satoh, M., Sonobe, K., Ishii, M., Ohtani, R., Kanamori-Sato, M., Honoki, R., Miyazaki, D., Mochizuki, H., Umetsu, J., Higashi, K., Shibata, D., Kamiya, Y., Sato, N., Nakamura, Y., Tabata, S., Ida, S., Kurokawa, K. and Ohta, H. (2014). *Klebsormidium flaccidum* genome reveals primary factors for plant terrestrial adaptation, *Nature Commun.*, 5, 3978.

Horrer, D., Flütsch, S., Pazmino, D., Matthews, J. S. A., Thalmann, M., Nigro, A., Leonhardt, N., Lawson, T. and Santelia, D. (2016). Blue light induces a distinct starch degradation pathway in guard cells for stomatal opening, *Curr. Biol.*, 26, pp. 362–370.

Howe, C. J., Barbrook, A. C., Nisbet, R. E. R., Lockhart, P. J. and Larkum, A. W. D. (2008). The origin of plastids, *Philos. Trans. R. Soc. Lond., B*, 363, pp. 2675–2685.

Hrda, Š., Fousek, J., Szabová, J., Hampl, V. V. and Vlček, Č. (2012). The plastid genome of *Eutreptiella* provides a window into the process of secondary endosymbiosis of plastid in Euglenids, *PLoS ONE*, 7, e33746.

Hu, H., Rappel, W.-J., Occhipinti, R., Ries, A., Böhmer, M., You, L., Xiao, C., Engineer, C. B., Boron, W. F. and Schroeder, J. I. (2015). Distinct cellular locations of carbonic anhydrases mediate carbon dioxide control of stomatal movements, *Plant Physiol.*, 169, pp. 1168–1178.

Jackson, C., Clayden, S. and Reyes-Prieto, A. (2015). The Glaucophyta: The blue-green plants in a nutshell, *Acta Soc. Bot. Pol.*, 84, pp. 149–165.

Janouškovec, J., Horák, A., Oborník, M., Lukeš, J. and Keeling, P. J. (2010). A common red algal origin of the apicomplexan, dinoflagellate, and heterokont plastids, *Proc. Natl. Acad. Sci. U.S.A.*, 107, pp. 10949–10954.

Jeffree, C. (1996). Structure and ontogeny of plant cuticles. *In* Kerstiens, G., ed, *Plant Cuticles: An Integrated Functional Approach.* (Bios Scientific Publishers, Oxford), pp. 33–82.

Jeon, S., Jeong, B. and Chang, Y. K. (2017). Chemicals and Fuels from Microalgae. *In* Lee, S. Y., ed, *Consequences of Microbial Interactions with Hydrocarbons, Oils, and Lipids: Production of Fuels and Chemicals.* (Springer International Publishing, Cham), pp. 1–21.

John, C. R., Smith-Unna, R. D., Woodfield, H., Covshoff, S. and Hibberd, J. M. (2014) Evolutionary convergence of cell-specific gene expression in independent lineages of C4 grasses, *Plant Physiol.*, 165, pp. 62–75.

Kamikawa, R., Tanifuji, G., Kawachi, M., Miyashita, H., Hashimoto, T. and Inagaki, Y. (2015). Plastid genome-based phylogeny pinpointed the origin of the green-colored plastid in the dinoflagellate *Lepidodinium chlorophorum*, *Genome Biol. Evol.*, 7, pp. 1133–1140.

Kandoi, D., Mohanty, S., Govindjee and Tripathy, B,C. (2016). Towards efficient photosynthesis: Overexpression of *Zea mays* phosphoenolpyruvate carboxylase in *Arabidopsis thaliana*, *Photosynth. Res.*, 130, pp 47–72.

Karki, S., Rizal, G. and Quick, W. P. (2013). Improvement of photosynthesis in rice (*Oryza sativa* L.) by inserting the C4 pathway, *Rice*, 6, 28.

Kebeish, R., Niessen, M., Thiruveedhi, K., Bari, R., Hirsch, H.-J., Rosenkranz, R., Stabler, N., Schonfeld, B., Kreuzaler, F. and Peterhansel, C. (2007). Chloroplastic photorespiratory bypass increases photosynthesis and biomass production in *Arabidopsis thaliana*, *Nat. Biotechnol.*, 25, pp. 593–599.

Keeling, P. J. (2013). The number, speed, and impact of plastid endosymbioses in eukaryotic evolution, *Annu. Rev. Plant Biol.*, 64, pp. 583–607.

Kim, S. and Park, M. G. (2016). *Paulinella longichromatophora* sp. nov., a new marine photosynthetic testate amoeba containing a chromatophore, *Protist*, 167, pp. 1–12.

Kim, T.-H., Böhmer, M., Hu, H., Nishimura, N. and Schroeder, J. I. (2010). Guard cell signal transduction network: Advances in understanding abscisic acid, CO_2, and Ca^{2+} signaling, *Annu. Rev. Plant Biol.*, 61, pp. 561–591.

Kirst, H., Gabilly, S. T., Niyogi, K. K., Lemaux, P. G. and Melis, A. (2017). Photosynthetic antenna engineering to improve crop yields, *Planta*, 245, pp. 1009–1020.

Kitzing, C. and Karsten, U. (2015). Effects of UV radiation on optimum quantum yield and sunscreen contents in members of the genera *Interfilum*, *Klebsormidium*, *Hormidiella* and *Entransia* (Klebsormidiophyceae, Streptophyta), *Eur. J. Phycol.*, 50, pp. 279–287.

Kneip, C., Voß, C., Lockhart, P. J. and Maier, U. G. (2008). The cyanobacterial endosymbiont of the unicellular algae *Rhopalodia gibba* shows reductive genome evolution, *BMC Evol. Biol.*, 8, pp. 30.

Krings, M., Kerp, H., Taylor, T. N. and Taylor, E. L. (2003). How Paleozoic vines and lianas got off the ground: On scrambling and climbing carboniferous-early permian pteridosperms, *Bot. Rev.*, 69, pp. 204–224.

Kromdijk, J., Głowacka, K., Leonelli, L., Gabilly, S. T., Iwai, M., Niyogi, K. K. and Long, S. P. (2016). Improving photosynthesis and crop productivity by accelerating recovery from photoprotection, *Science*, 354, pp. 857–861.

Krupke, A., Mohr, W., LaRoche, J., Fuchs, B. M., Amann, R. I. and Kuypers, M. M. M. (2015). The effect of nutrients on carbon and nitrogen fixation by the UCYN-A-haptophyte symbiosis, *ISME J.*, 9, pp. 1635–1647.

Kumar, A., Ogita, S. and Yau, Y.-Y., eds (2018). *Biofuels: Greenhouse Gas Mitigation and Global Warming*, (Springer India, New Delhi).

Larkum, A. W. D., Lockhart, P. J. and Howe, C. J. (2007). Shopping for plastids, *Trends Plant. Sci.*, 12, pp. 189–195.

Lauterborn, R. (1895) Protozoenstudien. *Z. Wiss. Zool.* 59, 537–544.

Leung, J., Bazihizina, N., Mancuso, S. and Valon, C. (2016). Revisiting the Plant's Dilemma, *Mol. Plant*, 9, pp. 7–9.

Li, F.-W., Rothfels, C. J., Melkonian, M., Villarreal, J. C., Stevenson, D. W., Graham, S. W., Wong, G. K. S., Mathews, S. and Pryer, K. M. (2015). The origin and evolution of phototropins, *Front. Plant Sci.*, 6, 637.

Lin, M. T., Occhialini, A., Andralojc, J. P., Devonshire, J., Hines, K. M., Parry, M. A. J. and Hanson, M. R. (2014). β-carboxysomal proteins assemble into highly organized structures in Nicotiana chloroplasts, *Plant J.*, 79, pp. 1–12.

Lu, Y. and Xu, J. (2015). Phytohormones in microalgae: A new opportunity for microalgal biotechnology?, *Trends Plant. Sci.*, 20, pp. 273–282.

Lubitz, W., Reijerse, E. J. and Messinger, J. (2008). Solar water-splitting into H_2 and O_2: Design principles of photosystem II and hydrogenases, *Energy Environ. Sci.*, 1, pp. 15–31.

Maier, A., Fahnenstich, H., von Caemmerer, S., Engqvist, M. K., Weber, A. P. and Flugge, U. I. (2012). Transgenic introduction of a glycolate oxidative cycle into A. thaliana chloroplasts leads to growth improvement, *Front Plant Sci.*, 3, 38.

Marshall, D. M., Muhaidat, R., Brown, N. J., Liu, Z., Stanley, S., Griffiths, H., Sage, R. F. and Hibberd, J. M. (2007). *Cleome*, a genus closely related to Arabidopsis, contains species spanning a developmental progression from C_3 to C_4 photosynthesis, *Plant J.*, 51, pp. 886–896.

Martin, B. A. and Frymier, P. D. (2017). A Review of hydrogen production by photosynthetic organisms using whole-cell and cell-free systems, *Appl. Biochem. Biotechnol.*, 183, pp. 503–519.

McAdam, S. A. M. and Brodribb, T. J. (2012). Fern and lycophyte guard cells do not respond to endogenous abscisic acid, *Plant Cell*, 24, pp. 1510–1521.

McGrath, J. M. and Long, S. P. (2014). Can the cyanobacterial carbon-concentrating mechanism increase photosynthesis in crop species? A theoretical analysis, *Plant Physiol.*, 164, pp. 2247–2261.

Merced, A. and Renzaglia, K. (2014). Developmental changes in guard cell wall structure and pectin composition in the moss *Funaria*: Implications for function and evolution of stomata, *Ann. Bot.*, 114, pp. 1001–1010.

Mikkelsen, M. D., Harholt, J., Ulvskov, P., Johansen, I. E., Fangel, J. U., Doblin, M. S., Bacic, A. and Willats, W. G. T. (2014). Evidence for land plant cell wall biosynthetic mechanisms in charophyte green algae, *Ann. Bot.*, 114, pp. 1217–1236.

Mullineaux, C. W. (2005). Function and evolution of grana, *Trends Plant. Sci.*, 10, pp. 521–525.

Murata, Y., Mori, I. C. and Munemasa, S. (2015). Diverse stomatal signaling and the signal integration mechanism, *Annu. Rev. Plant Biol.*, 66, pp. 369–392.

Nakayama, T. and Archibald, J. M. (2012). Evolving a photosynthetic organelle, *BMC Biol.*, 10, 35.

Nakayama, T., Ikegami, Y., Nakayama, T., Ishida, K.-I., Inagaki, Y. and Inouye, I. (2011). Spheroid bodies in rhopalodiacean diatoms were derived from a single endosymbiotic cyanobacterium, *J. Plant Res.*, 124, pp. 93–97.

Nakayama, T. and Inagaki, Y. (2014). Unique genome evolution in an intracellular N_2-fixing symbiont of a rhopalodiacean diatom, *Acta Soc. Bot. Pol.*, 83, pp. 409–413.

Nakayama, T., Kamikawa, R., Tanifuji, G., Kashiyama, Y., Ohkouchi, N., Archibald, J. M. and Inagaki, Y. (2014). Complete genome of a nonphotosynthetic cyanobacterium in a diatom reveals recent adaptations to an intracellular lifestyle, *Proc. Natl. Acad. Sci. U.S.A.*, 111, pp. 11407–11412.

Nevo, R., Charuvi, D., Tsabari, O. and Reich, Z. (2012). Composition, architecture and dynamics of the photosynthetic apparatus in higher plants, *Plant J.*, 70, pp. 157–176.

Nowack, E. C. M. (2014). *Paulinella chromatophora* – Rethinking the transition from endosymbiont to organelle, *Acta Soc. Bot. Pol.*, 83, pp. 387–397.

Ochoa de Alda, J. A. G., Esteban, R., Diago, M. L. and Houmard, J. (2014). The plastid ancestor originated among one of the major cyanobacterial lineages, *Nature Commun.*, 5, 4937.

Oey, M., Sawyer, A. L., Ross, I. L. and Hankamer, B. (2016). Challenges and opportunities for hydrogen production from microalgae, *Plant Biotechnol. J.*, 14, pp. 1487–1499.

Ort, D. R., Merchant, S. S., Alric, J., Barkan, A., Blankenship, R. E., Bock, R., Croce, R., Hanson, M. R., Hibberd, J. M., Long, S. P., Moore, T. A., Moroney, J., Niyogi, K. K., Parry, M. A. J., Peralta-Yahya, P. P., Prince, R. C., Redding, K. E., Spalding, M. H., van Wijk, K. J., Vermaas, W. F. J., von Caemmerer, S., Weber, A. P. M., Yeates, T. O., Yuan, J. S. and Guang Zhu, X. (2015). Redesigning photosynthesis to sustainably meet global food and bioenergy demand, *Proc. Natl. Acad. Sci. U.S.A.*, 112, pp. 8529–8536.

Parfrey, L. W., Lahr, D. J. G., Knoll, A. H. and Katz, L. A. (2011). Estimating the timing of early eukaryotic diversification with multigene molecular clocks, *Proc. Natl. Acad. Sci. U.S.A.*, 108, pp. 13624–13629.

Park, Y.-I., Choi, S.-B. and Liu, J. R. (2009). Transgenic plants with cyanobacterial genes, *Plant Biotechnol. Rep.*, 3, 267–275.

Parry, M. A. J., Keys, A. J., Madgwick, P. J., Carmo-Silva, A. E. and Andralojc, P. J. (2008). Rubisco regulation: A role for inhibitors, *J. Exp. Bot.*, 59, pp. 1569–1580.

Pfundel, E. and Pfeffer, M. (1997). Modification of photosystem I light harvesting of bundle-sheath chloroplasts occurred during the evolution of NADP-malic enzyme C4 photosynthesis, *Plant Physiol.*, 114, pp. 145–152.

Pierce, J. (1989). Rubisco: Mechanisms and their possible constraints on substrate specificity. *In* Briggs, W. R., ed, *Photosynthesis*. (Alan R. Liss, New York), pp. 149–159.

Portis, A. R. (2003). Rubisco activase – Rubisco's catalytic chaperone, *Photosynth. Res.*, 75, pp. 11 27.

Pribil, M., Labs, M. and Leister, D. (2014). Structure and dynamics of thylakoids in land plants, *J. Exp. Bot.*, 65, pp. 1955–1972.

Psenicka, J. and Oplustil, S. (2013). The epiphytic plants in the fossil record and its example from *in situ* tuff from Pennsylvanian of Radnice Basin (Czech Republic), *Bull. Geosci.*, 88, pp. 401–416.

Raschke, K. (1976). A discussion on water relations of plants - How stomata resolve the dilemma of opposing priorities, *Phil. Trans. R. Soc. Lond. B,* 273, pp. 551–560.

Roessler, P. G. and Lien, S. (1984). Activation and *de novo* synthesis of hydrogenase in *Chlamydomonas, Plant Physiol.,* 76, pp. 1086–1089.

Robinson, S. P. and Portis, A. R. (1989). Ribulose-1,5-bisphosphate carboxylase/oxygenase activase protein prevents the *in vitro* decline in activity of ribulose-1,5-bisphosphate carboxylase/oxygenase, *Plant Physiol.,* 90, pp. 968–971.

Romanowska, E., Drożak, A., Pokorska, B., Shiell, B. J. and Michalski, W. P. (2006). Organization and activity of photosystems in the mesophyll and bundle sheath chloroplasts of maize, *J. Plant Physiol.,* 163, pp. 607–618.

Rozema, J., Björn, L. O., Bornman, J. F., Gaberščik, A., Häder, D.-P., Trošt, T. and Germ, M. (2002). The role of UV-B radiation in aquatic and terrestrial ecosystems—an experimental and functional analysis of the evolution of UV-absorbing compounds, *J. Photochem. Photobiol. B: Biol.,* 66, pp. 2–12.

Rögner, M., ed (2015). *Biohydrogen.* (Walter de Gruyer GmbH, Berlin).

Savir, Y., Noor, E., Milo, R. and Tlusty, T. (2010). Cross-species analysis traces adaptation of Rubisco toward optimality in a low-dimensional landscape, *Proc. Natl. Acad. Sci. U.S.A.,* 107, pp. 3475–3480.

Sayre, R. (2010). Microalgae: The potential for carbon capture, *BioScience,* 60, pp. 722–727.

Seibert, M. and Torzillo, G., eds. (2018). *Microalgal Hydrogen Production: Achievements and Perspectives,* (RSC, Croydon).

Ševčíková, T., Horák, A., Klimeš, V., Zbránková, V., Demir-Hilton, E., Sudek, S., Jenkins, J., Schmutz, J., Přibyl, P., Fousek, J., Vlček, Č., Lang, B. F., Oborník, M., Worden, A. Z. and Eliáš, M. (2015). Updating algal evolutionary relationships through plastid genome sequencing: Did alveolate plastids emerge through endosymbiosis of an ochrophyte?, *Sci. Rep.,* 5, 10134.

Sharma, A. and Arya, S. K. (2017). Hydrogen from algal biomass: A review of production process, *Biotechnol. Rep.,* 15, pp. 63–69.

Sharwood, R. E. (2017a). A step forward to building an algal pyrenoid in higher plants, *New Phytol.,* 214, pp. 496–499.

Sharwood, R. E. (2017b). Engineering chloroplasts to improve Rubisco catalysis: Prospects for translating improvements into food and fiber crops, *New Phytol.,* 213, pp. 494–510.

Shih, P. M. and Matzke, N. J. (2013). Primary endosymbiosis events date to the later Proterozoic with cross-calibrated phylogenetic dating of duplicated ATPase proteins, *Proc. Natl. Acad. Sci. U.S.A.,* 110, pp. 12355–12360.

Shukla, V., Upreti, D. K. and Bajpai, R. (2014). *Lichens to Biomonitor the Environment.* (Springer, Berlin).

Stiller, J. W. (2003). Weighing the evidence for a single origin of plastids, *J. Phycol.*, 39, pp. 1283–1285.

Soda, N., Gupta, B. K., Anwar, K., Sharan, A., Govindjee, Singla-Pareek, S. L. and Pareek, A. (2018). Rice intermediate filament, OsIF, stabilizes photosynthetic machinery and yield under salinity and heat stress, *Sci. Rep.*, 8, 4072.

Sun, C. N. (1963). Submicroscopic structure and development of chloroplasts of *Marchantia polymorpha, J. Electron. Microsc.*, 12, pp. 254–259.

Suzuki, S., Shirato, S., Hirakawa, Y. and Ishida, K.-I. (2015). Nucleomorph genome sequences of two Chlorarachniophytes, *Amorphochlora amoebiformis* and *Lotharella vacuolata, Genome Biol. Evol.*, 7, pp. 1533–1545.

Tamagnini, P., Leitao, E., Oliveira, P., Ferreira, D., Pinto, F., Harris, D. J., Heidorn, T. and Lindblad, P. (2007). Cyanobacterial hydrogenases: Diversity, regulation and applications, *FEMS Microbiol. Rev.*, 31, pp. 692–720.

Taniguchi, Y., Ohkawa, H., Masumoto, C., Fukuda, T., Tamai, T., Lee, K., Sudoh, S., Tsuchida, H., Sasaki, H., Fukayama, H. and Miyao, M. (2008). Overproduction of C4 photosynthetic enzymes in transgenic rice plants: An approach to introduce the C4-like photosynthetic pathway into rice, *J. Exp. Bot.*, 59, pp. 1799–1809.

Thapper, A., Mamedov, F., Mokvist, F., Hammarström, L. and Styring, S. (2009). Defining the far-red limit of photosystem II in spinach, *Plant Cell*, 21, pp. 2391–2401.

Thompson, A., Carter, B. J., Turk-Kubo, K., Malfatti, F., Azam, F. and Zehr, J. P. (2014). Genetic diversity of the unicellular nitrogen-fixing cyanobacteria UCYN-A and its prymnesiophyte host, *Environ. Microbiol.*, 16, pp. 3238–3249.

Thompson, A. W., Foster, R. A., Krupke, A., Carter, B. J., Musat, N., Vaulot, D., Kuypers, M. M. M. and Zehr, J. P. (2012). Unicellular cyanobacterium symbiotic with a single-celled eukaryotic alga, *Science*, 337, pp. 1546–1550.

Tomitani, A., Okada, K., Miyashita, H., Matthijs, H. C. P., Ohno, T. and Tanaka, A. (1999). Chlorophyll *b* and phycobilins in the common ancestor of cyanobacteria and chloroplasts, *Nature*, 400, pp. 159–162.

Trissl, H. W. and Wilhelm, C. (1993). Why do thylakoid membranes from higher plants form grana stacks?, *Trends Biochem. Sci.*, 18, pp. 415–419.

Uehara, S., Adachi, F., Ito-Inaba, Y. and Inaba, T. (2016). Specific and efficient targeting of cyanobacterial bicarbonate transporters to the inner envelope membrane of chloroplasts in *Arabidopsis, Front. Plant Sci.*, 7, 16.

Vouilloud, A. A., Leonardi, P. I. and Cáceres, E. J. (2015). Mixed evolutionary traits of *Tolypella* (section Rothia, Charales) compared with *Chara* and *Nitella* shown by ultrastructure of vegetative internodal cells, *Aquat. Bot.*, 120, pp. 67–72.

Wang, C., Liu, Y., Li, S.-S. and Han, G.-Z. (2015). Insights into the origin and evolution of the plant hormone signaling machinery, *Plant Physiol.*, 167, pp. 872–886.

Xin, C.-P., Tholen, D., Devloo, V. and Zhu, X.-G. (2015). The benefits of photorespiratory bypasses: How can they work?, *Plant Physiol.*, 167, pp. 574–585.

Yoon, H. S., Hackett, J. D., Ciniglia, C., Pinto, G. and Bhattacharya, D. (2004). A Molecular timeline for the origin of photosynthetic eukaryotes, *Mol. Biol. Evol.*, 21, pp. 809–818.

Zannoni, D. and De Philippis, R., eds (2014). *Microbial BioEnergy: Hydrogen Production*, (Springer, Dordrecht).

Zehr, J. P., Shilova, I. N., Farnelid, H. M., del Carmen Muñoz-Marín, M. and Turk-Kubo, K. A. (2016). Unusual marine unicellular symbiosis with the nitrogen-fixing cyanobacterium UCYN-A, *Nature Microbiol.* 2, 16214.

Chapter 9

The Ultimate: Artificial Photosynthesis

9.1 Objectives and Approaches

In Chapter 8, we gave some examples of how scientists are attempting to improve photosynthesis (or rather adapt it for our needs) by changing specific genes by genetic engineering. But others have gone beyond this approach and are attempting to design artificial systems that carry out certain aspects of photosynthesis. These have different goals: production of electricity, fuels (including hydrogen), or chemicals such as plastics or pharmaceuticals, but not food. Another goal is to remove CO_2 from the atmosphere in order to decrease the greenhouse effect. Some of the constructs have been completely artificial, but others have incorporated components from the natural photosynthesis. There have been discussions on the theoretical efficiency of conversion of solar radiation to useful forms of energy. For example, Kleidon *et al.* [2016] have stated that the efficiency can be higher if photochemistry is used than if only the heat from the light is converted to other forms of energy. They have calculated an efficiency of 93% for direct solar radiation and 73% for completely scattered solar radiation, assuming an environmental temperature of 288 K and a solar spectrum as coming from a 5760 K blackbody.

In view of threatening global climate changes, and the prediction of general environmental disaster, much has been, and is being, invested in finding renewable energy sources. Rabinowitch [1961] had already discussed various ways of utilizing sunlight for storing energy through chemical reactions of dyes, by what was called the "photogalvanic" effect, but this did not go far because of extensive "back reactions." We note that research on

Photosynthesis: Solar Energy for Life by Dmitry Shevela, Lars Olof Björn and Govindjee
© 2018, published by World Scientific Publishing Co. Pte. Ltd. ISBN: 978-981-3223-10-3.

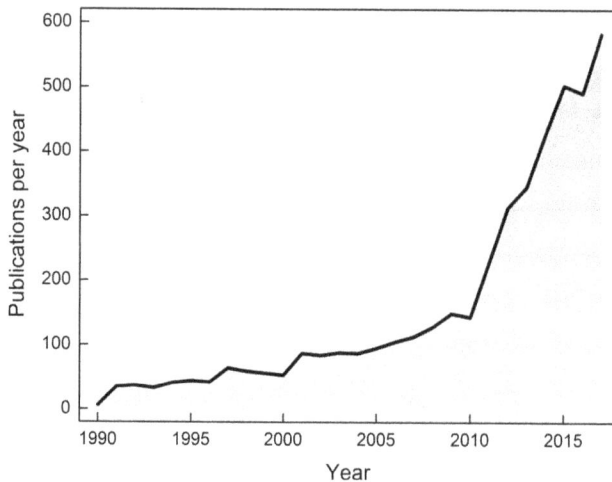

Fig. 9.1. Publications per year for the search string "artificial photosynthesis" as retrieved from the extended Web of Science.

artificial photosynthesis has increased dramatically over the past 25 years: the number of publications per year in 2017 was an order of magnitude higher than in 1991, as retrieved by searching for "artificial photosynthesis," in the Web of Science (see Fig. 9.1).

In view of the rapidly expanding literature, we shall provide here only some examples of research trends. For those who wish to delve deeper in this area, we refer them to several books [Collings and Critchley, 2005; Razeghifard, 2013; Robert *et al.*, 2016; de Vriend and Purchase, 2017].

9.2 Water Oxidation Coupled to Hydrogen Production: in Principle Simple, in Practice Not So Easy

A great deal of basic research has been concerned with the oxidation of water and the generation of usable reducing equivalents, a process that in the natural organisms takes place in and on thylakoid membranes (see Sections 2.3 [Chapter 2] and 3.3 [Chapter 3] and references therein; also see reviews by Lubitz *et al.* [2008]; Najafpour and Govindjee [2011]; Cox *et al.* [2015]; Barber [2017]). Naively we could think that this is very easy to achieve artificially: Just hook together some photovoltaic cells that generate current in the presence of light, with an electrolytic cell that splits water into molecular oxygen and hydrogen. Molecular hydrogen is, in some respects, an ideal, ecologically "clean" fuel. It is very light and energy compact (important for vehicles), and produces only water upon combustion. However, the main disadvantage is that it is difficult to store it in a safe way (for an example, see Hwang and Varma [2014]).

We must be cautious and critical. If one would set up this system without thinking further, we may end up with a contraption that may be hopelessly inefficient, and, in all likelihood, stop working in a very short period of time. For one thing, the older conventional photovoltaic cells were not very efficient, but some recent designs have shown energy efficiency exceeding 30% [Sivaram *et al.*, 2015; Green *et al.*, 2016]. The best so-called perovskite[*] cells have reached a little over 20% efficiency [Bella *et al.*, 2016; Saliba *et al.*, 2016; Tsai *et al.*, 2018]; this seems remarkable since they were only 5% efficient in 2010; thus, further improvement is expected. On the other hand, the best silicon cells have stayed at ~25% efficiency since the past century! We note that the efficiency spectrum for silicon cells peaks in the near infrared, while most perovskite cells do not function in the red and near infrared light, but they are efficient in blue light.

Thus, by putting perovskite cells on the top of silicon cells, we should be able to use sunlight much more efficiently and get a much higher total efficiency. Since specially designed perovskite cells absorb red light, all-perovskite devices, using both red and blue light, can now be constructed [Service, 2016] (*this reminds us of chlorophyll molecules*; see Chapter 2, Fig. 2.6). In addition, conversion of ultraviolet radiation (from the sunlight) to visible light can further boost performance [Bella *et al.*, 2016]. Since as a standard, efficiencies are reported for a "standardized" sunlight, corresponding to a solar zenith angle of 60°, efficiencies at specific wavelengths can be higher than these standardized values. One advantage of

---

[*] Perovskite refers to a crystalline material of the general composition ABX_3, where X is usually a halogen anion [Wang *et al.*, 2017], A is often an organic cation, and B is a metal cation in an octahedral lattice.

perovskite cells is their potentially low cost, which is partly due to the possibilities of making the manufacturing process highly automatic [Liu *et al.*, 2017]. However, there are still some issues with perovskites; by comparison, silicon cells are long-lived and have good resistance to various environmental factors, such as oxygen, humidity, heat, and light. However, for perovskite, Wang *et al.* [2017] have discussed ways to increase stability by choosing appropriate cations and halogen anions. On the other hand, Shin *et al.* [2017] have reported increased stability by the use of lanthanum-doped $BaSnO_3$ perovskite. Recently, Xu *et al.* [2018] developed an ultraflexible organic photovoltaic with sufficient thermal stability up to 120°C.

Another major reason to further demand improvement of the current methods is that in order to produce oxygen and hydrogen from water by electrolysis, we need a voltage much higher than what one predicts from thermodynamic considerations only. Thus, the equilibrium between water and hydrogen ($H_2O \leftrightarrow H_2 + 1/2\ O_2$) is characterized by a standard Gibbs free energy change (ΔG_0) of +237.2 kJ/mol [Renger, 2011], which corresponds to an electrical potential difference of 1.23 V at pH 7. However, in practice a higher voltage is required for this reaction, i.e., an "overvoltage" is needed, which results in energy loss in the form of heat [Mazloomi and Sulaiman, 2012]. Further, it is also necessary to add an electrolyte to the water to make it conductive. One might think that a good way would be to use seawater, with its natural high content of sodium chloride and other salts. However, the chloride ions are more easily discharged than the hydroxyl ions, so one would end up with Cl_2 instead of O_2. In the industrial electrolysis of water, potassium hydroxide (KOH) or sodium hydroxide (NaOH) is used at a high temperature, and at least 2 volts are applied at that temperature. Electrodes are made of metals that are resistant to high temperature as well as alkaline pH. However, we must remember that having high temperature requires an extra input of energy; thus, there are obvious problems to overcome in order to "generate" more energy than we would "use."

There is another thought. Instead of converting solar energy to electrical energy in one device, and then generating hydrogen by electrolysis of water in a separate unit, it may be more efficient and may be cost-effective, if we integrate the two processes in one unit. One such approach has been presented by Honda *et al.* [1969] and Fujishima and Honda [1972]; these authors have introduced the use of titanium dioxide, TiO_2, an n-type semi-conductor with a bandgap of 3 eV. They have constructed a cell with one electrode of TiO_2 and another one of platinum (see Fig. 9.2). When the TiO_2 electrode was irradiated with <415 nm light, O_2 was released at the TiO_2 electrode and H_2 was given off at the Pt-electrode. The quantum yield of the process was ~0.1, and since the cell does not use light with wave-lengths longer than 415 nm, the system is inefficient in daylight. However, since theoretically water can be decomposed in an electrolytic cell with a potential difference between the electrodes of 1.23 eV, one should, in prin-ciple, be able to construct a similar cell that could work with visible and near infrared radiation up to 1000 nm, a wavelength corresponding to a photon energy of 1.23 eV.

The work of Fujishima and Honda [1972] has been well recognized around the world: it had over 1800 citations in 2017, 45 years after its pub-lication, indicating that it is a pioneering and popular research finding; this implies that it may have initiated intensive research in many labs. In order to make it possible to exploit light over a wider spectrum, we would have to add to the original cell a suitable sensitizer which would absorb (capture) most of the available radiation. This, of course, requires further research.

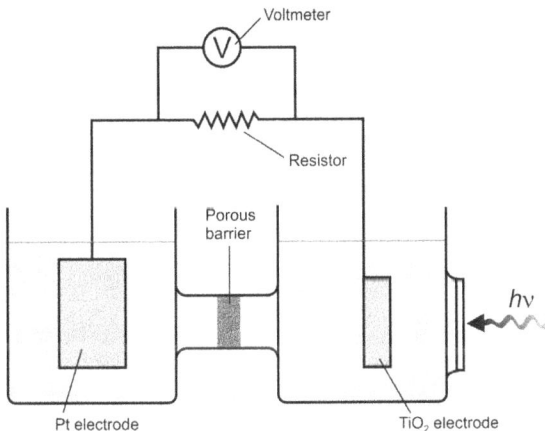

Fig. 9.2. The photoelectrochemical cell of Fujishima and Honda [1972]. A porous barrier is required to keep the evolved O_2 and H_2 separated. The cell can, in principle, develop a voltage of 0.5 V.

Since 1972, a large part of the periodic system of elements has been examined for use in research on artificial photosynthesis. For construction of a practically useful system, one must take into account not only its one-time efficiency, but also its deterioration with time, availability of the substances required, as well as the environmental risks associated with their large-scale use. From a quick look at the vast literature, it is not evident that the latter has always been in focus. Some of the metals (cadmium [Cd], lead [Pb], cobalt [Co] and nickel [Ni]) used thus far are dangerous for human health. Obviously, attempts have been (and are being) made to overcome this problem. Bock *et al.* [1974] and others observed light-induced electron transfer from Tris (2,2′-bipyridine) ruthenium (II) to specific electron acceptors when O'Regan and Grätzel [1991] had introduced a ruthenium (Ru) complex as a sensitizer for TiO_2; now, numerous modifications of this have been described (see, e.g., Brady *et al.* [2017]). As is evident from Fig. 9.3, the pioneering work of O'Regan and Grätzel has resulted in a remarkable extension of the spectral range for this system. Further, some newer work has provided remarkable ruthenium-based artificial antenna complexes. An example is shown in Fig. 9.4.

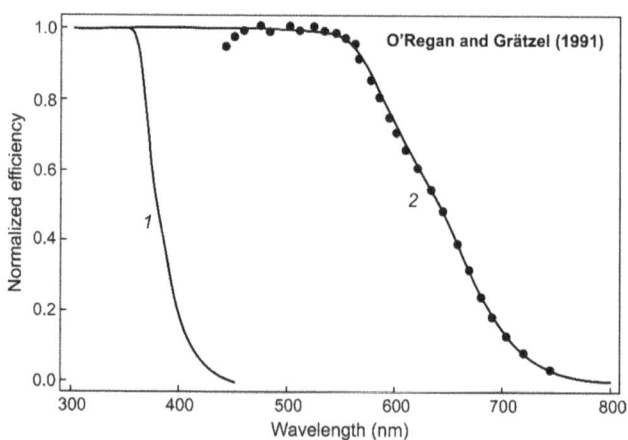

Fig. 9.3. Absorption and efficiency spectra for TiO_2 (*1*) and for TiO_2–Ru complex combination (*2*). The solid line *2* is for the absorption spectra, and the dots are for the efficiency. Modified from O'Regan and Grätzel [1991].

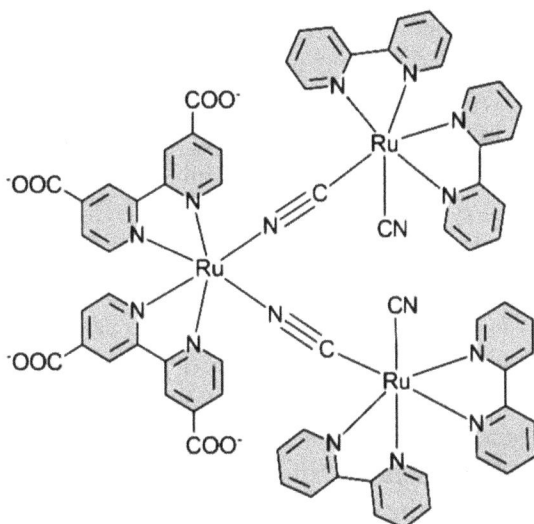

Fig. 9.4. Structure of one of many different kinds of artificial antennas based on ruthenium. Simplified from Pashaei *et al.* [2016].

The above-mentioned ruthenium-based system was described to be of "low cost." However, the price of ruthenium currently ranges from $2-5 per *g* (more than thousand times the price of TiO_2), which is really of "high cost." Now, alternatives are being searched for this system. A cheaper sensitizer complex, based on iron, has been described by Harlang *et al.* [2015]. This system generates, upon exposure to light, electrons in the conduction band of TiO_2 with a quantum yield of 0.92. However, it absorbs efficiently only up to about 550 nm; thus for efficient utilization of sunlight, we have a long way to go!

Another method of channeling light energy into TiO_2 is to use *quantum dots*. They are nano-sized semiconductor particles, which, because of their smaller size, in comparison to the wavelength of light, have special optical properties. Most of these constructs are being made by combining cadmium sulfide with several other compounds [Balis *et al.*, 2013; Samadpour, 2017]. A lot of effort has also been put in the construction of suitable counter-electrodes for such complex photoanodes [Balis *et al.*, 2013; Samadpour and Arabzade, 2017]! Here, also, we have a long way to go.

Furthermore, artificial antenna systems are now being constructed by using artificial polypeptides or proteins to which natural or artificial chromophores have been anchored, and that too, at carefully designed positions so as to allow efficient excitation energy transfer [Razeghifard and Wydrzynski, 2003; Chen *et al.*, 2005; Razeghifard *et al.*, 2007; Mancini *et al.*, 2017; Kodali *et al.*, 2017]. In our view, it is important to learn "ways and means" from the existing efficient natural systems for incorporation in future artificial systems (see, e.g., a review by Croce and van Amerongen [2014]). This approach would provide great flexibility in tuning, and is, thus, expected to allow the addition of artificial reaction centers [Wydrzynski *et al.*, 2007]. Work on such reaction centers, including artificial oxygen evolving centers, is also under way [Hirahara *et al.*, 2013; Najafpour *et al.*, 2013; Zhang *et al.*, 2015; Ye *et al.*, 2018].

As mentioned above, toxic metals have been used in some of the devices described above. It appears to us that concerns and solutions for the environmental consequences of the large-scale deployment of these systems need to be examined fully. Such research has begun to appear in the literature leading not only to proposals for metal-free alternatives [Markad *et al.*, 2017], but also for ways to recycle the available components [Huang *et al.*, 2017].

9.3 Reduction of Carbon Dioxide

If we can generate hydrogen from water, we can employ some organisms to use it as a reductant for assimilating CO_2 (see, e.g., Liu *et al.* [2016]; also see the website of Daniel Nocera: http://nocera.harvard.edu/Home). On the other hand, attempts are now being made not to use organisms for this purpose, but to carry out the reduction of carbon dioxide artificially. Of course no direct formation of carbohydrate has been (or can be) obtained. A number of ways in which CO_2 can be reduced is shown below, along with the required mid-point potentials:

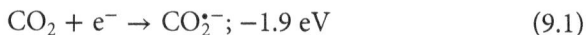

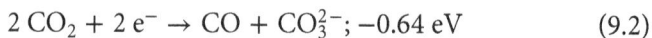

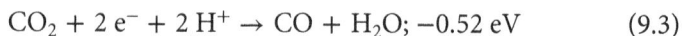

$$CO_2 + e^- \rightarrow CO_2^{\cdot -}; -1.9 \text{ eV} \tag{9.1}$$

$$2\,CO_2 + 2\,e^- \rightarrow CO + CO_3^{2-}; -0.64 \text{ eV} \tag{9.2}$$

$$CO_2 + 2\,e^- + 2\,H^+ \rightarrow CO + H_2O; -0.52 \text{ eV} \tag{9.3}$$

$$CO_2 + 2\,e^- + 2\,H^+ \rightarrow HCOOH; -0.61 \text{ eV} \tag{9.4}$$

The first, one-electron photoreduction (eq. 9.1), has thus far been realized only at the expense of an organic substance as an electron donor [Matsuoka *et al.*, 1992], and is thus of little interest from the viewpoint of CO_2 reduction, unless combined with a system that re-reduces the electron donor. Most systems described, thus far, convert CO_2 to either CO or to HCOOH. Neither of these options is suitable as a fuel for the vehicles because of toxicity or corrosivity. However, we note that CO was used for cars in Sweden during the Second World War, when the gasoline was not available! On the negative side, it caused 1135 cases of poisoning (in 1942), of which 11 were fatal (http://arkivgavleborg.blogspot.se/2014/05/gengas-fiende-eller-van.html). Further, at the present stage of our knowledge, this system seems unsuitable for another reason: it gives very little energy in comparison with the ensuing CO_2 emission. A typical hydrocarbon releases 11 to 12 kJ/g CO_2 emitted at combustion, while CO releases less than 5 kJ/g per molecule of CO_2 formed.

In the past, various transition metal complexes have been used as catalysts for CO_2 photoreduction [Yamazaki *et al.*, 2015]. However, recently catalysts employing graphene as one of the promising components have attracted attention [Yang and Xu, 2016].

Park *et al.* [2016] have obtained methane, mixed with carbon monoxide (CO), by illumination of TiO_2 combined with CdS in the presence of a mixture of water, isopropanol, and CO_2; here, in this system, isopropanol served as an electron donor. However, it turned out that only a fraction of the carbon in the methane originated from CO_2, and the major part came from isopropanol. Further, we know that CdS is a toxic substance, and, thus, care must be exercised for any use of this system.

Until now, the most successful systems have been those that use microorganisms for the assimilation of carbon dioxide, but in which hydrogen is obtained from electrolysis, or where extra energy is directly supplied electrically. An example of such a system is that of Liu *et al.* [2016] who used a suitable electrolysis cell with a cobalt-phosphate anode, and a NiMoZn alloy cathode; for the electrolyte, they used a bacterial growth medium. In this system, an external voltage of 3 V was applied, and *Ralstonia eutropha*, a bacterium, was grown in the cell, which ultimately converted the hydrogen released into an organic material, by using bicarbonate ions from the medium. We note that this organic material included a large proportion of alcohols that could indeed be used as fuel! Further, according to Liu *et al.* [2016], the energy efficiency was ~50%, and this was in the form of 180 g CO_2 assimilated per kWh. In combination with 25% efficient silicon solar cells and appropriate voltage matching, the overall efficiency of this system is expected to be 12.5%, which compares very well with any pure biological system. Of course, there will be a further loss of energy, and also emission of CO_2 when using the fuel, but the advantage with the mixture, used here, is that it is safer to store and use than hydrogen. The NiMoZn alloy, used for the cathode, seems to have been a great choice, partly because it minimized the formation of hydrogen peroxide, which is toxic to the bacterium.

Nichols *et al.* [2015] have described yet another hybrid system (Fig. 9.5) in which methanogens (belonging to Archaea), in an electrolytic half-cell,

$$2 H_2O + CO_2 \longrightarrow CH_4 + 2 O_2$$

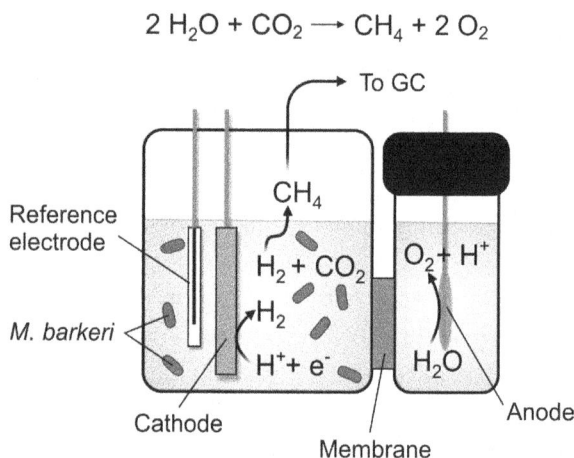

Fig. 9.5. Device generating methane and oxygen separately from carbon dioxide and water, using electrical energy and *Methanosarcina barkeri*, an archaeon. The notation "To GC" (see the above diagram) refers to the path of methane to a gas chromatograph, which is used to analyze it. The reference electrode serves to monitor pH changes. This basic setup can be modified in such a way that light first passes through a semitransparent anode of nanowires consisting of titanium dioxide on fluorine-doped tin oxide. The titanium dioxide is of n-type, i.e., a semiconducting crystal, with a slight excess of Ti over the stoichiometry in the formula TiO_2. After the light passes through a glass filter that removes blue light (to protect the organisms that would be damaged by blue light), it hits the cathode, which is made of indium phosphate coated with platinum. If the anode is connected to the cathode by a metal wire, a current flows through the circuit without an additional energy input and methane and oxygen are released separately. Modified from Nichols *et al.* [2015].

reduce carbon dioxide to methane by using electrolytically generated H_2, while oxygen is released in another half-cell, which is separated by a membrane from the methane-producing cell (Fig. 9.5). In these experiments, the rate of methane production was shown to be constant for more than 3 days. In the organism used, *Methanosarcina barkeri*, it appears that the same hydrogenase is used for the initial CO_2-reduction step as for the final step that leads to CH_4 formation [Krzyski and Zeikus, 1984].

The system, used by Liu *et al.* [2016], discussed above, is a hybrid system, with an artificial generation of a reductant (H_2) and a biological system that uses it to make products. *Ralstonia eutropha*, the organism used for CO_2 assimilation, uses the Calvin-Benson cycle to fix CO_2 (Chapter 4; Section 4.1). However, according to Boyle and Morgan [2011], the reductive acetyl-CoA pathway, the 3-hydroxypropionate/4-hydroxybutyrate cycle and the dicarboxylate/4-hydroxybutyrate cycle (Chapter 4; Section 4.5), are all energetically much more efficient than the Calvin-Benson cycle. Thus, there is room for further improvement here, by not using RuBisCO, which has a low affinity for carbon dioxide, has an affinity for oxygen, has a slow turnover, and is a resource-consuming complicated construct [Bracher *et al.*, 2017]. Bar-Even *et al.* [2010] suggest that artificial cycles be based on using natural enzymes combined in novel ways. Although two of their proposed cycles toward this goal use the efficient C4 enzyme, PEP carboxylase, yet the system ended up with the production of glyoxylate, which is not directly useful as a fuel. Thus, some modification is needed and further experiments must be done.

Asadi *et al.* [2016] have described what they call an "artificial leaf," where they use two photovoltaic cells in series acting as photosystem I and photosystem II (Fig. 9.6). To us, it has a future.

CO is certainly an unacceptable consumer product, both because of its toxicity and its small energy delivery on combustion (~283 kJ/mol). However, CO can be further processed, e.g., by electro-reduction on a suitably structured copper cathode to methane, and that, too, with a large energy content (~890 kJ/mol), or to ethylene for use in chemical industry. In fact, CO_2 can also be directly reduced to these chemicals on a copper cathode, but because CO_2 is converted to bicarbonate and carbonate at high pH, this process is limited to low pH. Further, it is also possible to electrolytically

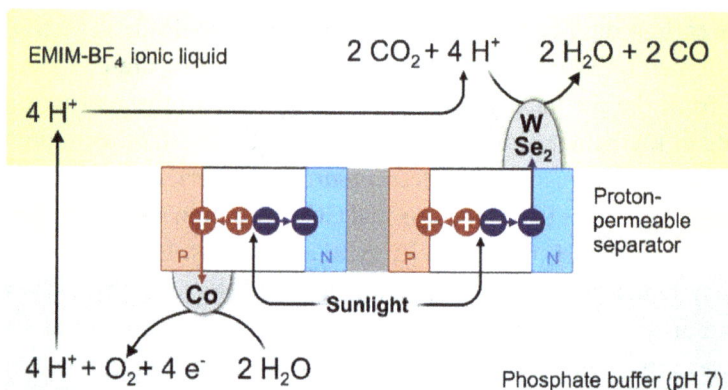

Fig. 9.6. A schematic representation of an "Artificial leaf," based on that by Asadi *et al.* [2016]. In the center are two photovoltaic cells connected in series by a metal connector, which is isolated from the liquids on both sides of the photovoltaic cells. The positive pole of this couple is connected to an electrode consisting of Co^{II}-oxide/hydroxide that is immersed in a pH 7 phosphate buffer. The negative pole is connected to an electrode consisting of tungsten diselenide (WSe_2), which is immersed in a special ionic liquid. The latter consists of 1-ethyl-3-methylimidazolium tetrafluoroborate (EMIM-BF4). Protons can move through the proton-permeable separator from the phosphate buffer to the EMIM-BF4, preventing a major change of pH in the phosphate solution. At the Co^{II}-oxide/hydroxide electrode, water is oxidized to O_2, while at the WSe_2 electrode, CO_2 is reduced to CO.

decompose, with high energetic efficiency, CO_2 dissolved in water to CO and O_2 by carefully designing the electrodes, which would avoid electrolysis of the water [Schreier *et al.*, 2017]. In addition to the above possibilities, CO can be combined with hydrogen (generated from water in a separate electrolytic cell) to form methane or other organic compounds. Thus, there is hope for the future!

9.4 Conclusions

Attempts to construct efficient and cheap artificial photosynthesis started out with attempts to imitate the ways plants perform photosynthesis, but it appears that the way forward may follow a very different course. A tremendous improvement of our understanding of Nature's photosynthesis, made possible during the last several years, provides us useful blueprints for the design of artificial photosynthetic machineries. These new discoveries in biological photosynthesis continue to inspire more and more research groups in the search for new ways to efficiently employ solar energy for fuel production [Barber and Tran, 2013; Razeghifard, 2013; Su and Vayssieres, 2016; Cogdell *et al.*, 2018]. It is interesting and impressive to mention that Lester P. Dutton and his associates (see, Kodali *et al.* [2017]; and Mancini *et al.* [2017]) are currently designing artificial systems based on in-depth knowledge of chemistry and physics, and the hope is to have a more efficient system, which may be better than what we have today. For a review on solar hydrogen technologies, see Ardo *et al.* [2018]. We are also keeping a watch on the research of many including young scientists (see e.g., Gary Moore website: https://sms.asu.edu/gary_moore).

We end this chapter and our book by citing George P. Shultz, an American economist, elder statesman and a businessman, who said: "*The source of all the energy is the sun. The big challenge is, how do you use all of that energy? Solar power has to fascinate you. There have been strides to get the costs down, and if this will work, you have to get costs down so it is competitive with fossil fuels*". All the three authors of this book have high hopes for the future.

References

Ardo, S., Rivas, D. F., Modestino, M., Greiving, V. S., Abdi, F., Llado, E. A., Artero, V., Ayers, K., Battaglia, C., Becker, J-P., Bederak, D., Berger, A., Buda, F., Chinello, E., Dam, B., Palma, V. D., Edvinsson, T., Fujii, K., Gardeniers, H., Geerlings, H., Hashemi, M., Haussener, S., Houle, F., Huskens, J., James, B., Konrad, K., Kudo, A., Kunturu, P. P., Lohse, D., Mei, B., Miller, E., Moore, G. F., Muller, J., Orchard, K., Post, R., Rosser, T., Saadi, F., Schüttauf, J-F., Seger, B., Sheehan, S., Spurgeon, J., Tang, M., van de Krol, R., Vesborg, P. and Westerik, P. (2018). Pathways to electrochemical solar hydrogen technologies. *Energy Environ. Sci.*, DOI:10.1039/c7ee03639f

Asadi, M., Kim, K., Liu, C., Addepalli, A. V., Abbasi, P., Yasaei, P., Phillips, P., Behranginia, A., Cerrato, J. M., Haasch, R., Zapol, P., Kumar, B., Klie, R. F., Abiade, J., Curtiss, L. A. and Salehi-Khojin, A. (2016). Nanostructured transition metal dichalcogenide electrocatalysts for CO_2; reduction in ionic liquid, *Science*, 353, pp. 467–470.

Balis, N., Dracopoulos, V., Bourikas, K. and Lianos, P. (2013). Quantum dot sensitized solar cells based on an optimized combination of ZnS, CdS and CdSe with CoS and CuS counter electrodes, *Electrochim. Acta*, 91, pp. 246–252.

Bar-Even, A., Noor, E., Lewis, N. E. and Milo, R. (2010). Design and analysis of synthetic carbon fixation pathways, *Proc. Natl. Acad. Sci. U.S.A.*, 107, pp. 8889–8894.

Barber, J. (2017). Photosynthetic water splitting by the $Mn_4Ca^{2+}O_x$ catalyst of photosystem II: Its structure, robustness and mechanism, *Q. Rev. Biophys.*, 50, p. e13.

Barber, J. and Tran, P. D. (2013). From natural to artificial photosynthesis, *J. R. Soc. Interface*, 10, p. 20120984.

Bella, F., Griffini, G., Correa-Baena, J.-P., Saracco, G., Grätzel, M., Hagfeldt, A., Turri, S. and Gerbaldi, C. (2016). Improving efficiency and stability of perovskite solar cells with photocurable fluoropolymers, *Science*, 354, pp. 203–206.

Bock, C. R., Meyer, T. J. and Whitten, D. G. (1974). Electron transfer quenching of the luminescent excited state of tris(2,2′-bipyridine)ruthenium(II). Flash photolysis relaxation technique for measuring the rates of very rapid electron transfer reactions, *J. Am. Chem. Soc.*, 96, pp. 4710–4712.

Boyle, N. R. and Morgan, J. A. (2011). Computation of metabolic fluxes and efficiencies for biological carbon dioxide fixation, *Metab. Eng.*, 13, pp. 150–158.

Bracher, A. Whitney, S. M., Hartl, F. U. and Manajit Hayer-Hartl, M. (2017). Biogenesis and metabolic maintenance of Rubisco, *Annu. Rev. Plant Biol.*, 68, pp. 29–60.

Brady, M. D., Sampaio, R. N., Wang, D., Meyer, T. J. and Meyer, G. J. (2017). Dye-sensitized hydrobromic acid splitting for hydrogen solar fuel production, *J. Am. Chem. Soc.*, 139, pp. 15612–15615.

Chen, M., Eggink, L. L., Hoober, J. K. and Larkum, A. W. D. (2005). Influence of structure on binding of chlorophylls to peptide ligands, *J. Am. Chem. Soc.*, 127, pp. 2052–2053.

Cogdell, R. J., Gardiner, A. T., Yukihira, N., Hashimoto, H. (2018). Solar fuels and inspiration from photosynthesis, *J. Photochem. Photobiol. A Chem.*, 353, pp. 645–653.

Collings, A. F. and Critchley, C., eds, (2005). *Artificial Photosynthesis: From Basic Biology to Industrial Application* (Willey-VCH, Weinheim).

Cox, N., Pantazis, D. A., Neese, F. and Lubitz, W. (2015). Artificial photosynthesis: Understanding water splitting in nature, *Interface Focus*, 5, p. 20150009.

Croce, R. and van Amerongen, H. (2014). Natural strategies for photosynthetic light harvesting, *Nat. Chem. Biol.*, 10, pp. 492–501.

de Vriend, H. and Purchase, R. (2017). Solar Fuels and Artificial Photosynthesis: Science and Innovation to Change Our Future Energy Option. Based on a Report by The Royal Society of Chemistry (BioSolar Cells, Wageningen, The Netherlands).

Fujishima, A. and Honda, K. (1972). Electrochemical photolysis of water at a semiconductor electrode, *Nature*, 238, pp. 37–38.

Green, M. A., Emery, K., Hishikawa, Y., Warta, W. and Dunlop, E. D. (2016). Solar cell efficiency tables (version 47), *Prog. Photovolt. Res. Appl.*, 24, pp. 3–11.

Harlang, T. C. B., Liu, Y., Gordivska, O., Fredin, L. A., Ponseca Jr, C. S., Huang, P., Chábera, P., Kjaer, K. S., Mateos, H., Uhlig, J., Lomoth, R., Wallenberg, R., Styring, S., Persson, P., Sundström, V. and Wärnmark, K. (2015). Iron sensitizer converts light to electrons with 92% yield, *Nat. Chem.*, 7, pp. 883–889.

Hirahara, M., Shoji, A. And Yagi, M. (2013). Artificial manganese center models for photosynthetic oxygen evolution in photosystem II, *Eur. J. Inorg. Chem.*, 2014, pp. 595–606.

Honda, K., Fujishima, A. and Kikuchi, S. (1969). Photosensitized electrolytic oxidation on semiconducting n-type TiO_2 electrode, *J. Chem. Soc. Jpn.*, 72, pp. 108–113.

Huang, W.-H., Shin, W. J., Wang, L., Sun, W.-C. and Tao, M. (2017). Strategy and technology to recycle wafer-silicon solar modules, *Sol. Energy*, 144, pp. 22–31.

Hwang, H. T. and Varma, A. (2014). Hydrogen storage for fuel cell vehicles, *Curr. Opin. Chem. Eng.*, 2014, 5, pp. 42–48.

Kleidon, A., Miller, L. and Gans, F. (2016). Physical limits of solar energy conversion in the earth system. *In* Tüysüz, H., Chan, C. K., eds, *Solar Energy for Fuels. Topics in Current Chemistry*, Vol. 371 (Springer, Cham), pp. 1–22.

Kodali, G. Mancini, J. A., Solomon, L. A., Episova, T. V., Roach, N., Hobbs, C. J., Wagner, P., Mass, O. A., Aravindu, K., Barnsley, J. E., Gordon, K. C., Officer, D. L., Dutton, P. L. and Moser, C. C. (2017). Design and engineering of water-soluble light-harvesting protein maquettes, *Chem. Sci.*, 8, pp. 316–324.

Krzyski, J. A. and Zeikus, J. G. (1984). Characterization and purification of carbon monoxide dehydrogenase from *Methanosarcina barkeri*, *J. Bacteriol.*, 158, pp. 231–237.

Liu, C., Colón, B. C., Ziesack, M., Silver, P. A. and Nocera, D. G. (2016). Water splitting–biosynthetic system with CO_2 reduction efficiencies exceeding photosynthesis, *Science*, 352, pp. 1210–1213.

Liu, S., Cao, K., Li, H., Song, J., Han, J., Shen, Y. and Wang, M. (2017). Full printable perovskite solar cells based on mesoscopic $TiO_2/Al_2O_3/NiO$ (carbon nanotubes) architecture, *Sol. Energy*, 144, pp. 158–165.

Lubitz, W., Reijerse, E. J. and Messinger, J. (2008). Solar water-splitting into H_2 and O_2: Design principles of photosystem II and hydrogenases, *Energy Environ. Sci.*, 1, pp. 15–31.

Mancini, J.-A., Kodali, G., Jiang, J., Reddy, K. R., Lindsey, J. S., Bryant, D. A., Dutton, P. L. and Moser, C. C. (2017). Multi-step excitation energy transfer engineered in genetic fusions of natural and synthetic light-harvesting proteins, *J. R. Soc. Interface*, 14, p. 20160896.

179

Markad, G. B., Kapoor, S., Haram, S. K. and Thakur, P. (2017). Metal free, carbon-TiO_2 based composites for the visible light photocatalysis, *Sol. Energy*, 144, pp. 127–133.

Mazloomi, S. K. and Sulaiman, N. (2012). Influencing factors of water electrolysis electrical efficiency, *Renew. Sustain. Energy Rev.*, 16, pp. 4257–4263.

Matsuoka, S., Kohzuki, T., Pac, C. Ishida, A. Takamuku, S., Kusaba, M., Nakashima, N. and Yanagida, S. (1992). Photocatalysis of oligo(p-phenylenes). Photochemical reduction of carbon dioxide with triethylamine, *J. Phys. Chem.*, 96, pp. 4437–4442.

Najafpour, M. M. and Govindjee (2011). Oxygen evolving complex in Photosystem II: Better than excellent, *Dalton Trans.*, 40, pp. 9076–9084.

Najafpour, M. M., Tabrizi, M. A., Behzad Haghighia, B. and Govindjee (2013). A 2-(2-hydroxyphenyl)-1H-benzimidazole–manganese oxide hybrid as a promising structural model for the tyrosine 161/histidine 190-manganese cluster in photosystem II, *Dalton Trans.*, 42, pp. 879–884.

Nichols, E. M., Gallagher, J. J., Liu, C., Su, Y., Resasco, J., Yu, Y., Sun, Y., Yang, P., Chang, M. C. Y. and Chang, C. J. (2015). Hybrid bioinorganic approach to solar-to-chemical conversion, *Proc. Natl. Acad. Sci. U.S.A.*, 112, pp. 11461–11466.

O'Regan, B. and Grätzel, M. (1991). A low-cost, high-efficiency solar cell based on dye-sensitized colloidal TiO_2 films, *Nature*, 353, pp. 737–740.

Park, H., Ou, H.-H., Kang, U., Choi, J. and Hoffmann, M. R. (2016). Photocatalytic conversion of carbon dioxide to methane on TiO_2/CdS in aqueous isopropanol solution, *Catal. Today*, 266, pp. 153–159.

Pashaei, B., Shahroosvand, H., Graetzel, M. and Nazeeruddin, M. K. (2016). Influence of ancillary ligands in dye-sensitized solar cells, *Chem. Rev.*, 116, pp. 9485–9564.

Rabinowitch, E. (1961). Photochemical utilization of light energy, *Proc. Natl. Acad. Sci. U.S.A.*, 47, pp. 1296–1303.

Razeghifard, R., ed, (2013). *Natural and Artificial Photosynthesis: Solar Power as an Energy Source* (John Wiley & Sons Inc., Hoboken, NJ).

Razeghifard, M. R. and Wydrzynski, T. (2003). Binding of Zn-chlorin to a synthetic four-helix bundle peptide through histidine ligation, *Biochemistry*, 42, pp. 1024–1030.

Razeghifard, M. R., Wallace, B. B., Pace, R. J. and Wydrzynski, T. (2007). Creating functional artificial proteins, *Curr. Protein Pept. Sci.*, 8, pp. 3–18.

Renger, G. (2011). Light induced oxidative water splitting in photosynthesis: Energetics, kinetics and mechanism, *J. Photochem. Photobiol. B*, 104, pp. 35–43.

Robert, B. ed. (2016). Artificial photosynthesis. *Adv. Bot. Res.*, 79, pp. 1–248.

Saliba, M., Matsui, T., Domanski, K., Seo, J.-Y., Ummadisingu, A., Zakeeruddin, S. M., Correa-Baena, J.-P., Tress, W. R., Abate, A., Hagfeldt, A. and Grätzel, M. (2016). Incorporation of rubidium cations into perovskite solar cells improves photovoltaic performance, *Science*, 354, pp. 206–209.

Samadpour, M. (2017). Efficient CdS/CdSe/ZnS quantum dot sensitized solar cells prepared by ZnS treatment from methanol solvent, *Sol. Energy*, 144, pp. 63–70.

Samadpour, M. and Arabzade, S. (2017). Graphene/CuS/PbS nanocomposite as an effective counter electrode for quantum dot sensitized solar cells, *J. Alloy Comp.*, 696, pp. 369–375.

Schreier, M., Héroguel, F., Steier, L., Ahmad, S., Luterbacher, J. S., Mayer, M. T., Luo, J. and Grätzel, M. (2017). Solar conversion of CO_2 to CO using Earth-abundant electrocatalysts prepared by atomic layer modification of CuO, *Nature Energy*, 2, p. 17087.

Service, R. F. (2016). Perovskite solar cells gear up to go commercial, *Science*, 354, pp. 1214–1215.

Shin, S. S., Yeom, E. J., Yang, W. S., Hur, S., Kim, M. G., Im, J., Seo, J., Noh, J. H. and Seok, S. (2017). Colloidally prepared La-doped $BaSnO_3$ electrodes for efficient, photostable perovskite solar cells, *Science*, 356, pp. 167–171.

Sivaram, V., Stranks, S. D. and Snaith, H. J. (2015). Outshining silicon, *Sci. Am.*, 313, pp. 54–59.

Su, J. and Vayssieres, L. (2016). A place in the Sun for artificial photosynthesis?, *ACS Energy Lett.*, 1, pp. 121–135.

Tsai, H., Asadpour, R., Blancon, J.-C., Stoumpos, C. C., Durand, O., Strzalka, J. W., Chen, B., Verduzeo, R., Ajayan, P. M., Tretiak, S., Even, J., Alam, M. A., Kanatzidis, M. G., Nie, W. and Mohite, A. D. (2018). Light-induced lattice expansion leads to high-efficiency perovskite solar cells, *Science*, 360, pp. 67–70.

Wang, Z., Shi, Z., Li, T., Chen, Y. and Huang, W. (2017). Stability of perovskite solar cells: A prospective on the substitution of the A-cation and X-anion, *Angew. Chem. Int. Ed.*, 56, pp. 1190–1212.

Wydrzynski, T., Hillier, W. and Conlan, B. (2007). Engineering model proteins for photosystem II function, *Photosynth. Res.*, 94, pp. 225–233.

Xu, X., Fukuda, K., Karki, A., Park, S., Kimura, H., Jinno, H., Watanabe, N., Yamamoto, S., Shimomura, S., Kitazawa, D., Yokota, T., Umezu, S. and Nguyen, T.-Q. (2018). Thermally stable, highly efficient, ultraflexible organic photovoltaics, *Proc. Natl. Acad. Sci. U.S.A.*, 115, pp. 4589–4594.

Yamazaki, Y., Takeda, H. and Ishitani, O. (2015). Photocatalytic reduction of CO_2 using metal complexes, *J. Photochem. Photobiol. C: Photochem. Rev.*, 25, pp. 106–137.

181

Yang, M.-Q. and Xu, Y.-J. (2016). Photocatalytic conversion of CO_2 over graphene-based composites: Current status and future perspective, *Nanoscale Horiz.*, 1, pp. 185–200.

Ye, S., Ding, C., Chen, R., Fan, F., Fu, P., Yin, H., Wang, X., Wang, Z., Du, P. and Li, C. (2018). Mimicking the key functions of photosystem II in artificial photosynthesis for photoelectrocatalytic water splitting, *J. Am. Chem. Soc.*, 140, pp. 3250–3256.

Zhang, C., Chen, C., Dong, H., Shen, J.-R., Dau, H. and Zhao, J. (2015). A synthetic Mn_4Ca-cluster mimicking the oxygen-evolving center of photosynthesis, *Science*, 348, pp. 690–693.

Index

187